T0227597

Routledge Introductions to Development
Series Editors
John Bale and David Drakakis-Smith

The Arab World

The Arab oil states have not, unlike many other parts of the world, been constrained primarily by a lack of capital, but by factors as diverse as water availability and shortages of skilled labour. This has produced distinctive and unusual development situations in the Arab region.

The Arab World contrasts Arab and western perspectives on the resources and economic potential of the region, arguing that external involvement has been responsible for many of the development dilemmas faced by the Arabs today. More importantly, the crisis of Arab identity has detracted from and often destroyed promising schemes for economic advancement; development objectives for an Arab future must satisfy both the rising economic aspirations and the strong attachment to Islamic values of the peoples of the region.

Questioning whether economic progress is possible without social and political change, this introductory text offers students of geography, economics, sociology and development studies a unique evaluation of the means by which states as diverse as Saudi Arabia, Yemen, Morocco and Jordan have sought to achieve economic advancement in relation to the cultural contexts from which they have emerged.

Allan M. Findlay is Senior Lecturer in Geography and Research Co-Ordinator of the Applied Population Research Unit, University of Glasgow.

In the same series

Janet Henshall Momsen
Women and Development in the Third World
David Drakakis-Smith
The Third World City
Allan and Anne Findlay
Population and Development in the Third World
Avijit Gupta
Ecology and Development in the Third World
John Lea
Tourism and Development in the Third World
John Soussan
Primary Resources and Energy in the Third World
Chris Dixon
Rural Development in the Third World
Alan Gilbert
Latin America
David Drakakis-Smith
Pacific Asia
Rajesh Chandra
Industrialization and Development in the Third World
Mike Parnwell
Population Movements and the Third World
Tony Binns
Tropical Africa
Jennifer A. Elliot
An Introduction to Sustainable Development
The Developing World

Forthcoming

George Cho
Global Interdependence
Trade, Aid, and Technology Transfer

Ronan Paddison
Retail Patterns in the Third World

Allan M. Findlay

The Arab World

Routledge
Taylor & Francis Group

LONDON AND NEW YORK

First published 1994
by Routledge

2 Park Square, Milton Park, Abingdon, Oxon OX14 4RN
711 Third Avenue, New York, NY 10017, USA

Routledge is an imprint of the Taylor & Francis Group, an informa business

First issued in hardback 2016

Typeset in Times by J&L Composition Ltd, Filey, North Yorkshire

Transferred to digital printing 2003

British Library Cataloguing in Publication Data

A catalogue record for this book is available from the British Library

Library of Congress Cataloging in Publication Data

Findlay, Allan M.
  The Arab world / Allan M. Findlay.
    p.   cm. – (Routledge introductions to development)
  Includes bibliographical references and index.
    1. Arab countries – Economic conditions.   I. Title.   II. Series.
  HC498.F56 1993
  330.917'6927 – dc20   93–5461

ISBN 978-0-415-04200-0 (pbk)
ISBN 978-1-138-16223-5 (hbk)

To my parents

Contents

List of plates   x
List of figures   xi
List of tables   xii
Preface   xiii
Acknowledgements   xvi

1 **Perspectives on Arab development: from West to East**   1
Introduction   1
The geographical division of the Arab world   1
Gross national product per capita and life expectancy   3
Perceptions of the physical environment   7
Perceptions of natural resources   9
Perceptions of Arab society   11
Case study A: Lebanon – a myriad of minorities   13
Perceptions of Arab politics and government   18
Conclusion   21
Key ideas   22

2 **The colonial legacy**   23
Introduction   23
The Ottoman influence   23
European influences   26
French colonization of the Maghreb   27
Case study B: 'Modernization' in Tunisia – the colonial
experience   28

The partitioning of the Middle East   34
Case study C: The House of Saud and the emergence of a
kingdom   38
Conclusion   43
Key ideas   43

3   Political constraints to economic development   **45**
Introduction   45
Religious and ethnic divisions   46
The cost of conflict   49
Palestine and Israel   52
Case study D: The Balfour Declaration   53
Case study E: Baqaa refugee camp   59
Refugees   65
Weak government   65
Conclusion   70
Key ideas   70

4   Arab oil and the use of oil revenues   **71**
Introduction   71
The organization of the oil industry in the Arab world   73
Who controls Arab oil?   74
Case study F: The growth and development of ARAMCO   85
Oil prices as an indicator of oil power   86
Infrastructure development in the oil states   88
The Libyan path to development in the 1970s and 1980s   91
Agricultural and industrial development in the Gulf states   93
Kuwait: the emergence of a rentier state   97
Development paths and development objectives   99
Conclusion   100
Key ideas   101

5   Labour migration   **102**
Introduction   102
Migration and development in labour-sending states   103
The shifting demand for immigrant labour   104
Housing and labour market implications of migration   110
Case study G: Yemen: the world's most migrant-dependent state   110
The use of remittances   118
International migration and settlement patterns   119
Case study H: Migration and the urban growth of Amman city   120

Changing times; changing migrants   123
Key ideas   125

6 Rural development   **126**
Introduction   126
Water: a critical resource   128
Dams and irrigation   132
Case study I: Development in the desert: Wadi Allaqi, Egypt   138
Capital, water and agriculture   142
Nomadism   143
Sedentarization   147
Land reform   149
Food security versus rural employment   153
Conclusion   158
Key ideas   159

7 Urban development   **160**
Introduction   160
Urbanism and urbanization in the Arab world   161
The Islamic city?   162
Images of the city   165
Case study J: A day in the life of a Cairo rubbish collector   178
Arab banking as an indicator of urban economic development   183
Conclusion: urban development and urban crises   188
Key ideas   189

8 Arab identity and development   **191**
Introduction   191
Progress without change?   193
Key idea   195

References, further reading and review questions   **196**
Index   **203**

Plates

1.1   The High Atlas mountain range, Morocco   7
3.1   Housing in Baqaa refugee camp, Jordan   60
3.2   Main street of Baqaa refugee camp   61
5.1   Craftsmen chewing *qat* in the jewellery *suq*, Sanaa   111
5.2   Fortress house, western highlands, Yemen   112
5.3   Speculative urban development on the outskirts of Amman, Jordan   121
6.1   Irrigation canal and agriculture, Fayoum Depression, Egypt   127
6.2   Olive plantations, the Sahel region of Tunisia   127
6.3   Threshing floor by a dry wadi in the eastern highlands of Yemen   128
6.4   Farmer with land enclosure in Wadi Allaqi, Egypt   138
7.1   The abandoned city of Marib, Yemen   166
7.2   *Medina* and colonial city, Rabat   167
7.3   One of the many *suqs* of the *medina* of Tunis   168
7.4   An example of the street layout in the colonial-built city – Avenue Habib Bourguiba, Tunis   169
7.5   Young refuse collectors, Cairo   179

Figures

1.1   Per capita gross national product and life expectancy at birth (1991)   4
1.2   Selected physical features of the Arab world   8
1.3   Arab peoples and selected ethnic minorities in the Arab world   12
1.4   Lebanon: too many fingers in the pie   15
2.1   Growth of the Ottoman empire   24
2.2   The evolution of the Tunisian rail network   29
2.3   Port hinterlands in the Tunisian Protectorate   30
2.4   European colonization and Arab independence   37
2.5   The emergence of the Saudi state   39
3.1   The branches of Islam   47
3.2   The world refugee crisis   64
4.1   World oil reserves, oil production and oil consumption, 1990   72
4.2   Who controls oil? A schematic representation of the world oil industry   76
4.3   The ups and downs of world oil prices, 1972–91   82
4.4   Provincial distribution of industrial licences in Saudi Arabia (1984) and the value of planned industrial investment   96
5.1   Amman's urban growth   120
6.1   Deviation in annual discharge for the Blue Nile at Khartoum (metres), 1900–89   131
6.2   Annual lake level deviation for Lake Victoria (metres), 1900–88   132
6.3   Lake Asad and associated Syrian irrigation schemes   134
6.4   Location of Wadi Allaqi   139
7.1   Squatter settlements of Cairo   182
7.2   Location of major Arab bank headquarters, 1980   184
8.1   Development without change?   192

Tables

1.1	A basic data matrix on the Arab region	2
2.1	Populations of coastal and interior cities of Morocco at the beginning and end of the colonial era	33
3.1	State budget and military expenditure in selected states of the Arab world in the 1980s	50
4.1	The Arab oil industry: some basic indicators for 1990	75
5.1	The world's most migrant dependent states	104
5.2	Estimated stock of immigrants in certain Arab oil states	105
5.3	Remittance use by return and current migrants in Jordan, 1984	119
5.4	Estimated numbers of selected migrant groups in Iraq and Kuwait, July 1990	124
6.1	The extension of irrigation in Morocco during the 1973–7 National Plan	136
6.2	Irrigation improvement targets, 1975–90	143
6.3	Calendar of activities in the oasis of Kurudjel, 1989–90	145
6.4	Indicators of the significance of agricultural systems	153
7.1	Modal number of hours worked per day	171
7.2	Official aid from Arab countries and agencies, 1973–89	175

Preface

For the past sixteen years it has been my privilege to carry out research on development issues in the Arab world. Initially my research took me to Tunisia and Morocco, but from the early 1980s onwards new opportunities emerged to work in Jordan, Yemen and Egypt. The longer I have worked in the region, the more I have become aware of the acute contrast between Western perceptions of the Arabs and the picture which one develops from actually living and working in Arab lands. This book commences by asking the reader to re-investigate his or her own images of the Arabs, prior to turning to any consideration of development issues in the region. It is only when one realizes that many Arabs hold very different views of development from those which are popular in the West that one can begin to understand why the Arab states have followed such divergent 'development' paths.

The book covers territory from Yemen in the east to Mauritania in the west, and from Syria and Iraq in the north to Sudan in the south. The focus is on development and how different forces have competed to mould the future of the Arab world. As a consequence the text does not attempt to provide the even spatial coverage of a traditional regional geography book (see for example Beaumont *et al.*, 1988; Chapman and Baker, 1992), but instead it strives to illustrate, from across the Arab region, those issues which are of most concern to students of development.

Chapters 2 and 3 show how the colonial era and continued Western intervention have contributed to the political and economic instability of the Arab world, and consequently have hindered its development. To the West oil is undoubtedly perceived as the single most important physical resource of the Arab world. Having discussed the relationship between Arab oil and the world oil industry (Chapter 4), analysis turns to the broader implications of oil revenues for the disparate development courses pursued by the Arab oil states. In the next chapter it is shown that the other Arab states have also been strongly affected by their proximity to the oil economies, through on the one hand the implications for their labour markets of emigration of large numbers of their active population, and on the other hand the impact on their economies of the in-flow of substantial foreign earnings in the form of remittances.

Chapter 6 centres attention on the rural economy and asks what the appropriate development path is for the Arab lands with their limited water resources and their rapidly expanding populations. It is in the urban arena, however, that one can most clearly perceive the struggle between different 'development' forces (Chapter 7). Urban space is contested by the many different influences which are moulding the future of the Arab world. The book concludes by suggesting that a problem remains for the Arabs in defining their own identity. Without such a definition there will continue to be 'change', but not necessarily 'progress'. These changes, whether they be in the rural or urban economies of the Arab states, will result from a suite of influences including the ongoing impact of the world economy, the attempts by some Arab leaders to oppose external influence and Western models of development, and the efforts of other Arab policy makers to pursue paths which they perceive to represent truly 'Arab' development.

Not only is this text unusual in seeking to include in one volume a discussion of development issues from both North Africa and the Arab Middle East, but the timing of its preparation, in coming just after the war in which the Iraqi army was expelled from Kuwait, means that development issues are viewed in a new historical setting. One could say that virtually all development issues in the Arab world, and certainly those discussed in Chapters 3–8 of this book, need to be considered afresh in the light of the changed position of Arab states relative to each other and to the world economy following the war.

My own perceptions of the Arab world owe much to the time and efforts of a host of Arab friends and colleagues who have taken time to introduce me over the past two decades to 'their' world. In particular I would like to thank Habib Abichou, Mohammed Maani, Bassam Nasr,

Musa Samha and Faisal Zanoun. I hope this book is not unfaithful to the goal of transferring at least some of this understanding to others. I hope they will forgive me for views expressed in this book which they do not share. Amongst British colleagues I am very grateful to Brian Beeley, Gerald Blake, Brian Clark, John Clarke and Richard Lawless who in different ways made it possible for me to carry out my early research in Arab lands. I also owe an immense debt of gratitude to my French colleagues Pierre Signoles, Jean-Marie Miossec and Jean Francois Troin, who have proved an immense encouragement at times when few British geographers seemed interested in the Arab world. John Briggs, John Jowett, Ronan Paddison and Ian Thompson at Glasgow University also played an important part in making this book possible. Finally my thanks must go to my parents and to my wife, Anne, who had much more to do with the completion of this book than I can ever adequately acknowledge.

Acknowledgements

I would like to thank the following for permission to use material in this book: John Jowett for Figure 1.1; I. Thompson for Plate 1.1; the International Labour Office, Geneva, for material used in Chapter 5; Michael Hulme and CRU, University of East Anglia, for Figures 6.1 and 6.2; John Briggs for Case study I, Figure 6.4 and Plate 6.4; Juliette Brown and Belhaven Publishers for Table 5.1; Croom Helm Publishers for Tables 5.3 and 6.1; Urbama, University of Tours, for Table 6.3, Figure 7.2, Plate 7.2 and material on Arab banking previously published in a different form in *Cahiers D'Urbama*, vol. 6; Gunter Meyer and *Applied Geography and Development* for Case study J and Plate 7.4. I am also most grateful to M. Shand, Y. Wilson and L. Hill who prepared the illustrations for this volume.

Perspectives on Arab development: from West to East

Introduction

Perceptions of economic and social development are often illusory or misleading, while measures of development often reflect more about the bias and perceptions of the investigator than about the lived circumstances of the people to whom they refer. Nowhere is this more true than for Westerners examining the Arab world. The goal of this chapter is to make the reader more aware of his or her own bias concerning development issues in the Arab world, before turning in later chapters to a consideration of specific dimensions of development in the Arab region. The chapter commences by considering the geographical units and some of the statistical indicators commonly employed in the analysis of development. It then focuses briefly on four dimensions of the Arab world (physical environment, natural resources, society and politics) and for each one seeks to contrast Western and Arab perceptions.

The geographical division of the Arab world

Briefly consider Table 1.1. It shows, amongst other features, that the citizens of some Arab states such as Saudi Arabia enjoy both considerable wealth (as measured in terms of gross national product (GNP) per capita) and the prospects of a moderately high life expectancy. People in other Arab countries such as Yemen are much less fortunate on both counts. But what do statements such as these indicate about Western as opposed to Arab perspectives on development?

Table 1.1 A basic data matrix on the Arab region[a]

State	Population (millions) 1991	Life expectancy at birth (years) 1991	GNP per capita (US$) 1989	Oil production[b] (% of total world volume) 1990	Urban population (%) 1991
Algeria	26.0	64	2,170	1.9	48
Egypt	54.5	57	630	1.4	45
Iraq	17.1	64	n.a.	3.1	73
Israel[c]	4.9	76	9,750	–	90
Jordan	3.4	71	1,730	–	70
Kuwait[d]	1.4	(73)	(14,870)	1.7	(95)
Lebanon	3.4	68	n.a.	–	80
Libya	4.4	67	5,410	2.1	70
Mauritania	2.1	47	490	–	41
Morocco	26.2	62	900	–	46
Oman	1.6	64	5,220	1.0	9
Saudi Arabia	15.5	63	6,230	10.4	73
Sudan	25.9	51	420	–	20
Tunisia	8.4	64	1,260	–	54
Syria	12.8	69	1,020	–	50
United Arab Emirates	2.4	71	18,430	3.5	78
Yemen	10.1	50	640	–	25

Sources: Population Reference Bureau, *1991 World Population Data Sheet*; British Petroleum, *BP Statistical Review of World Energy 1991*

Notes: [a] Excludes states with a population of less than 1 million (Bahrain, 0.5 million; Qatar, 0.5 million) and Occupied Territories (Western Sahara, 0.2 million; Gaza, 0.6 million; West Bank, 1.1 million).

[b] Production figures accounting for less than 1.0 per cent of the world total are excluded. Thus for example both Tunisia and Yemen produced some oil in 1990, but this is not recorded in this table.

[c] Excludes Gaza and the West Bank.

[d] Figures in parentheses refer to 1989–90 or the latest reported position before the Iraqi invasion of 1990.

n.a., not available.

The table, like many others in this book, has already imposed on the reader a Western conception of development. In geographical terms the table fragments the Arab world into territorial units described as 'states', rather than treating the Arab world as a whole or seeking to partition it in terms of how Arabs perceive their region. The state units used in Table 1.1 were mostly imposed on the Arabs by Western powers over the last 100 years. Both the boundaries of the states and the political structures which were established have been far from neutral influences on Arab development. The boundaries of the states were largely drawn up by Westerners in relation to external interests and perceptions of the resources (locational, military, natural and human) of the Arab territories.

The state of Iraq serves well to illustrate this point. Consider the following quotations which concern the creation of Iraq in 1930 and its subsequent 'development'.

[Before 1930] There never had been an independent Arab Iraq, nor was there any demand among the local population for an Iraqi state.

(Peretz, 1978, 406)

Not only did Iraq lack a convincing *raison d'être*, but . . . many of its early leaders sought to prevent the development of a distinct Iraqi identity, preferring to emphasize its Arab identity. . . . Nationalists assumed that Iraq would eventually disappear and be submerged in some larger Arab identity.

(Drysdale and Blake, 1985, 272)

We don't look on this piece of land, here in Iraq, as the ultimate limit of our struggle. It is part of a larger area and broader aims, the area of the Arab homeland and the aims of the Arab struggle. We look on our Iraqi people of twelve million as part of a people of 140 million. We look at the present divisions as an unnatural state which must be ended with unity.

(Sadam Hussein, President of Iraq, 1975)

These quotations serve to show how hard it is to measure development in the Arab world without imposing, almost unknowingly, Western concepts such as 'the state' upon one's analysis. If it is difficult to avoid this problem even at the level of the territorial units of study, how much harder it is when one starts to seek suitable indicators of economic and social development.

Gross national product per capita and life expectancy

Discussion of development in this chapter commences with considera-tion of the most widely used Western indicator of economic activity, the GNP per capita. This indicator measures the financial value of the total production of goods and services in a state relative to the population of the country. Although the measure is often used to estimate the average income in a country (i.e. by dividing the total value of production by the number of people) this measure is clearly flawed since in most

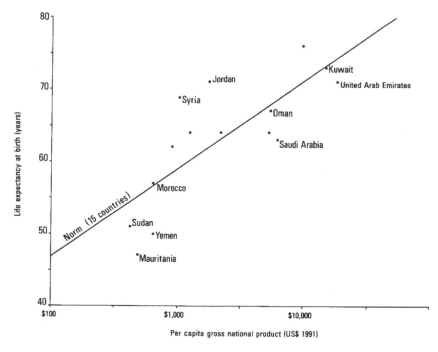

Figure 1.1 Per capita gross national product and life expectancy at birth (1991)
Source: A. Jowett, APRU, University of Glasgow

countries there is a highly inequitable distribution of the value of production between the inhabitants. GNP per capita is an indicator which makes at least two assumptions about Arab development that limit its usefulness: first, that development is in some way associated with the total value of economic production rather than with some other measure or interpretation of development, and second, that economic activities are organized in a fashion which can be measured in a meaningful quantitative sense in order to prepare national accounts.

Bearing these points in mind, what would one be able to conclude as an outsider examining the Arab world using this indicator? Figure 1.1 makes possible some interesting generalizations about the Arab world and about the differences between the Arab states and the states of the European Community (EC). The diagram shows that in those Arab states for which data are available there seems to be an association between national wealth, as measured by GNP per capita, and life expectancy at birth.

Examination of Figure 1.1 seems to suggest that life expectancies rise very rapidly with small increases in GNP per capita amongst the poorest countries and then much more slowly with further increases in wealth in the richest Arab countries. The development literature makes it clear that the relationship is a complex one and is by no means uni-directional. Amongst the many forces which might lead one to expect there to be a relationship are (a) the improved nutritional levels which populations experience as countries develop and as the increased wealth is distributed amongst the population and (b) the increased capability of the state to pay for and provide medical facilities and education, to tackle disease, and to increase life expectancies.

Life chances were worst in Mauritania and Yemen where a child in 1991 had a life expectancy at birth of 50 years or less. In Yemen the achievement of a life expectancy of 50 years was itself a big improvement on former circumstances, given that life expectancies of only 39 years at birth were recorded as recently as 1970 (Yemen Arab Republic). Sudan with an average income of only US$420 and a life expectancy of 51 years was little better. At the other end of the spectrum the peoples of the United Arab Emirates enjoyed life expectancies of 71 years and a GNP per capita of over US$18,000. Prior to the war with Iraq, Kuwait boasted life expectancies of 73 years and a per capita income of US$14,870. The diversity of circumstances in the Arab world can scarcely be over-emphasized.

The reader should note the split between the major Gulf oil and gas producing countries, all of which had GNPs per capita greater than US$5,000, and the much more populous and relatively poor non-oil states. The fact that the Arab world is divided in this way is underlined by the following statistics. The Gulf oil states and Libya (excluding Algeria and Iraq) all appear on Figure 1.1 with GNPs per capita of over US$5,000, yet they had in total a population of less than 27 million people in 1991, while the countries on the graph with incomes less than US$2,000 per capita had between them a population of approximately 143 million.

The first significant contrast which emerges from Figure 1.1 is then between the 'rich few' in the Arab world, many of whom have good life expectancies, and the 'poor majority' whose life chances are somewhat more varied depending on the economic history and political development strategies of the states in which they live. For example, Jordan falls into the poorer category of countries on the diagram yet life expectancy at birth was 71 years, as good as any of the oil states except

Kuwait. If one were to interpret life expectancy as an indicator of more than simply demographic circumstances and to accept it as a broader indicator reflecting one dimension of quality of life, then it would be fair to say in the Arab countries that rising standards of living do seem to have become associated with improvements in other aspects of their population's quality of life. Improved life expectancy and other aspects of the development of human potential has not, however, been limited only to the oil-rich states, but has also been experienced by people in countries as widely differing in character as Jordan, Tunisia and Syria.

A second important contrast is between the Arab and the West European states. While the twelve countries of the EC have almost as diverse a range of economic performances as the Arab states if this is measured in terms of GNP per capita, they are much more homogeneous in terms of the life expectancies of their populations. All of them can offer their people a life expectancy of between 74 and 77 years. The marginal differences in life expectancy which do exist do not seem to relate to variations in GNP per capita, suggesting that whatever might be the development forces accounting for further improvements in life expectancy, in countries which already offer very good life opportunities, they do not seem to stem directly from increases in the volume of economic activity. This might point to the conclusion that development needs to be measured in ways other than the purely economic, while in no way negating the relevance of economic change as one dimension of the development process.

Two final brief points should be noted from Figure 1.1. First it illustrates the fallacy of the popular Western perception that the Arabs are all wealthy. Not one of the EC states has a GNP per capita level as low as that experienced in the Arab states of the 143 million 'poor majority' defined above, and with the exception of Portugal, Greece, Spain and Ireland all have per capita GNP levels higher than Saudi Arabia. Second, they all boast life expectancies which are still significantly higher than those of the Arab states.

Not only do Western assessments of Arab development often rest on externally imposed criteria and peculiarly selective and partial indicators of development, but they also differ from Arab perceptions of their own region in terms of the potential for future development. This is evident whether one considers the physical environment or the economic, social or political dimensions of the Arab world.

Perceptions of the physical environment

Figure 1.2 presents some aspects of the physical geography of the Arab world. From a Western perspective it is often the vast deserts of the region which are the most striking feature. This is not without some justification since, as the map shows, most of the region receives very little rainfall and over most of the area evaporation of moisture into the atmosphere has the potential to exceed precipitation. Arab perspectives of the region are dominated, not by the deserts, but by the water resources made available by great rivers such as the Nile, Euphrates and Tigris which bring water into the area and which have been the focus of settlement patterns throughout most of Arab history. The heartlands of Arab civilizations have evolved here as well as in other areas of water surplus such as in the highlands of Yemen, the Levant and the northern part of the Maghreb. As Table 1.1 shows most Arabs are not desert pastoralists but city dwellers. In Syria 50 per cent of the population live in cities, in Jordan the proportion is 70 per cent while in Iraq it rises to 73 per cent.

Plate 1.1 The High Atlas mountain range, Morocco

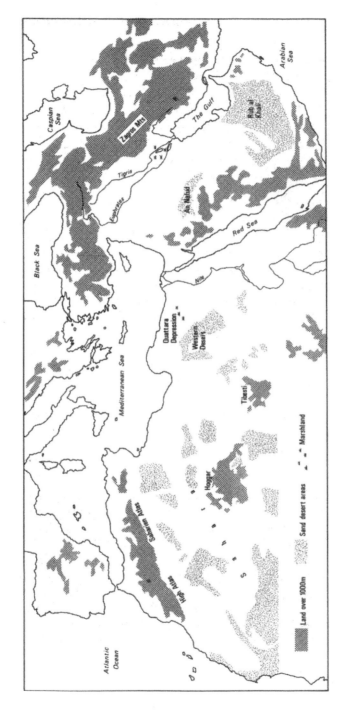

Figure 1.2 Selected physical features of the Arab world

To Western nations the physical structure of the Arab world has historically been perceived as a barrier to travel and trade with India and the east. Consequently Western concerns have often been to establish and secure sea routes around the Arabian peninsula. As well as constructing the Suez canal (which was opened in 1869 after 15 years of obstruction by the British) consideration was also given in the mid-nineteenth century to the idea of building a canal from Haifa to Aqaba. Even before the construction of the Suez canal, the overland transfer of goods via Suez and the Red Sea saved three months on a voyage from London to Calcutta. As has been amply demonstrated by the many confrontations in the latter half of the twentieth century between Western armed forces and the Arab powers, this difference of perspective has not diminished. The West has continued to be concerned about access to the Suez canal and the Red Sea, and also about control of the Straits of Hormuz in relation to access to the oilfields of the Gulf.

Arab power has usually been land based, relating for example, even as early as the times of King Solomon and the Queen of Sheba (Queen Bilqis), to control of the caravan routes across the desert. Throughout most of Arab history the great military leaders have been those with powerful land armies capable of controlling territory rather than concerned with strategic sea routes. Transport and communications for the Arabs has therefore been mainly over land rather than by sea, and prior to the present century this gave rise to the inland location of the dominant routeways and associated settlements.

Perceptions of natural resources

If one turns from the physical environment to consider Western and Arab perspectives on other matters such as natural resources, one again finds major differences. One economic issue above all others dominates Western thinking about Arab resources. It is oil. In the 1950s and 1960s Western Europe, Japan and to a lesser extent the United States depended very heavily on oil produced in the Middle East. These industrialized nations have never forgotten the way in which the Arab oil producers used the so-called 'oil weapon' in 1973 to bring about changes in the West's support for Israel. Perhaps it is because oil was used by the Arabs to promote a cause cherished both by the oil-rich and non-oil Arab states that it has often been perceived by the West as a major uniting force in the Arab region. In practice this has proved a false perception.

Although in 1990 the Middle East still accounted for 66 per cent of the world's known oil reserves (the United States had 3 per cent and Western Europe 2 per cent), oil remained a dominantly divisive force in the Arab countries. Not only was it a source of resentment between the oil-rich and oil-poor states of the region, but it was also a source of bitter dispute between the Arab oil producers. The soaring riches of the Arab oil states and the international power and prestige which oil was seen to give to the political leaders of the oil producing states, such as Saudi Arabia and Kuwait, challenged the position of the traditional leaders of the Arab world who have viewed cities like Cairo, Damascus and Baghdad as the cultural and political capitals of the Arabs. A parallel contest for prestige and power was evident in the 1970s and 1980s between oil-rich Libya and the states of the Maghreb, such as Morocco with its long traditions of urbanism and its potentially rich agricultural base.

Conflict between oil producers has arisen because their economic and political interests are so disparate. For example, a key factor lying behind the Iraqi invasion of Kuwait in August 1990 was the difference between these two states' perceptions of oil: Iraq, wishing to rebuild its economy after the long war with Iran, was acutely in need of foreign exchange and therefore eager to raise crude oil prices. By contrast Kuwait, with its large oil reserves and with its greater investment in the refining and retailing of petroleum, wanted to resist price rises which might have reduced its filling station sales and encouraged a new round of investment by Western oil companies in exploration and production activities outside the Arab world.

While the West has been concerned about securing Arab oil to fuel its energy-greedy industries and transport systems, oil is not found in many of the Arab countries and the natural resource which Arab states have spent the greatest effort to develop is water. The aridity of the region has meant that since earliest times rulers of the Arab countries have made great efforts to acquire new sources of water, to distribute it more efficiently and to recycle its use. Consider for example the wealth and degree of social organization which must have existed in Yemen 3,000 years ago to permit the construction of the Marib dam. This dam across Wadi Dhana is believed to have fed an irrigated agricultural area which was capable of feeding an estimated 30,000–40,000 people. The technology of the twentieth century combined with the wealth accumulated from oil revenues has made possible new and ambitious schemes to access and distribute underground water reserves.

Libya, for example, devoted very significant portions of its oil revenues in the 1980s to build what has become known as 'the great man-made river' to pump 5 million cubic metres of water per day from underground reservoirs in the south of the country to the coastal agricultural regions.

Since access to water is the key in nearly all of the region to the production of food crops, it is not surprising that in most rural communities the inheritance of water rights has been guarded as jealously as that of land. In the twentieth century this is as true at the level of the state as it has been for individual cultivators throughout the history of agricultural settlement in the region. When Arab leaders of the twentieth century have met, it has often been to discuss 'water security', but these meetings have seldom been reported in the West because, unlike oil, water resources have been perceived as an entirely Arab affair. In practice water security has not of course been an internal affair. Turkey has intermittently interrupted the flow of the Euphrates into Iraq in order to implement some of its rural development projects (which involve the building of no less than twenty-one dams across the headwaters of the river), while Israel has diverted much of the discharge of the river Jordan from Lake Tiberias into its national water network at the expense of Jordanian and Palestinian farmers.

Perceptions of Arab society

On social issues, the Western view of the Arabs is so strongly coloured by the contrasts in culture which arise from linguistic and religious differences with the West that internal diversity and social tensions are often forgotten (Figure 1.3). For example, in answering the question 'Who are the Arabs?' the well known Professor of Arabic at Oxford University, H. A. Gibb, stated in 1940 (prior to the establishment of many of today's modern independent Arab states):

> there is only one answer – whatever ethnographers may say – which approaches historic truth: all those are Arabs for whom the central fact of history is the mission of Mohammed and the memory of the Arab Empire, and who cherish the Arabic tongue and its cultural heritage as their common possession.
>
> (Gibb, 1940, 3)

The vision of re-creating one great Arab nation has certainly been a powerful influence in temporarily uniting some Arab states at times of

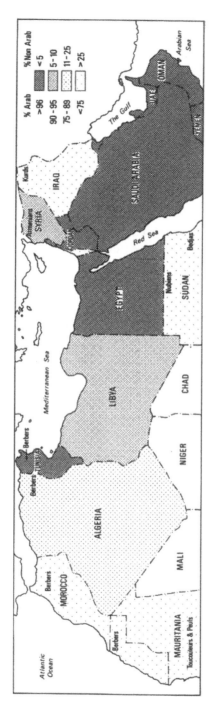

Figure 1.3 Arab peoples and selected ethnic minorities in the Arab world

conflict with the West, while the recent resurgence of Islamic fundamentalism, as represented most forcefully in 1979 by the revolution in
Iran (a non-Arab state), has proved to be a force little understood by
Western politicians and economic leaders. But in reality not all Arabs
are Muslim. Consider, for example, the substantial Christian Coptic
communities of Egypt which number 3.5 million persons, or alternatively
the many minority religious groups of Lebanon.

Case study A

Lebanon – a myriad of minorities

Lebanon was once described by Westerners as the Switzerland of
the Middle East. This description was not entirely inappropriate
prior to the outbreak of Lebanon's bitter civil war in 1975.
Lebanon has impressive mountain chains and an abundance of
water. Its capital, Beirut, became in the 1960s the banking capital
of the Arab world and a major centre for international trade
between the Arabs and the West. The most important aspect of
the analogy with Switzerland, however, was that Lebanon was
composed of a regional patchwork of distinctive ethnic and
cultural groups, yet unlike other parts of the Middle East these
minority groups seemed, to the eyes of the outsider at least, to co-
exist moderately harmoniously.

The present boundaries of Lebanon were established by the
French after the First World War. Within the boundaries of
Lebanon were at least six major groups. Maronite Christians were
the largest group and were to be found mainly in the valleys of
north and central Lebanon and in the coastal cities of Beirut,
Tripoli, Sidon and Tyre. Their military power was to become
established in the Phalangist militia. Sunni Muslims were the
dominant group to the north of Tripoli. Sunni Muslims also made
up the largest part of the populations of Beirut and Tripoli. In the
south of Lebanon and in the Beqaa valley in the east were Shi'ite
Muslims, described in Lebanon as Metwali and protected by their
own militia, later to be known as Amal. The Druze, although
having Muslim origins, are a sect rejected as true Muslims by most
Islamic peoples. They live in the Shouf mountains, southeast of
Beirut. In addition to these four large groupings the state of

Case study A (*continued*)

Lebanon also found itself host to significant minorities of Armenian Christians who had fled to Beirut following repeated persecution by the Turks in the 1890s and during the First World War (over 600,000 are believed to have been massacred at this time). Beirut also became home to about a quarter of a million Greek Orthodox Christians.

By the early 1990s no one was sure of the precise size of the different sects in Lebanon, since no census had been taken since 1932. It was clear, however, that the demographic balance of the country had been severely upset by substantial emigration from amongst the Christian groups and by the much higher fertility levels of the Muslim population. Both the Shi'ites and the Maronites claimed to represent over 1 million people. There were estimated to be about 750,000 Sunni Muslims and 200,000 Druze. The geographical effect of almost twenty years of bitter civil war in Lebanon has been to strengthen the spatial concentration of minorities in the different parts of Lebanon, with countless accusations being made by all sides of the 'ethnic cleansing' of isolated village communities.

Despite spells of peaceful co-existence Lebanon was never the Switzerland of the Middle East that Westerners wanted to perceive it to be. After independence in 1943 the role of central government within the state was a very weak one. It was based on a pact (the much referred to Lebanese National Pact) by which the Maronites renounced French protection (and implicitly other Western support), the Muslims agreed to give up the struggle to become part of a much larger pan-Arab state, and all parties agreed to 'confessionalism'. Confessionalism meant that the state would not interfere with the authority of local religious and clan leaders within each of the communities of Lebanon. Effectively each group was left to police its own sphere with its militia, to raise local taxes and customs duties and to enforce its own value systems.

The pact was inevitably fragile. When the Maronite President of Lebanon supported the West during the 1956 conflict with Nasser, other groups in Lebanon felt the pact had been violated, a feeling strengthened by the subsequent deployment of US troops to uphold the Maronite President. The result was Lebanon's first civil war in 1958.

Case study A (*continued*)

Figure 1.4 Lebanon: too many fingers in the pie

The system of a 'state within a state' which the National Pact had effectively instituted made Lebanon a natural location for

Case study A (*continued*)

other minorities to seek refuge. Some Palestinians had fled there after the creation of the state of Israel in 1947, but the main influx followed the defeat of the Palestinian struggle in Jordan in 1971. In the early 1970s south Lebanon became the refuge for several hundred thousand Palestinians and in particular became the most secure base available to the Palestinian Liberation Organization (PLO). Israel, unhappy about the PLO's new strength in southern Lebanon, began to intervene increasingly in the country's affairs, giving support to the Maronites to drive out the Palestinians. Once again the fragile truce between Lebanon's myriad of minorities was destabilized and civil war broke out. Syrian intervention in 1976 ironically supported the 'Christian' Phalangist forces, crushed for a time the various Muslim militias, and constrained the operations of the PLO. In 1982 Israel invaded the south of Lebanon to displace the Palestinian population there. This simply transferred the conflict to Beirut and led to a new struggle between the Palestinian groups supported by Syria and those led by Yasser Arafat's PLO. It also led to the subsequent intervention of the Syrian army in large numbers across much of Lebanon.

The civil war in Lebanon has never really come to an end. Instead fighting has persisted, often occurring within factions as well as between sects. In 1990 the Maronites experienced a bitter internal struggle between those willing to accept a new plan for the country (involving the Maronites having less power than before, but also guaranteeing the disarming of the militia and the withdrawal of Syrian troops) and those Maronites still wishing to return to the old order. By 1992 Syrian troops were still in Lebanon, the militias were still involved in periodic confrontations and the Maronites had refused to participate in elections to establish a new government. Ironically (perhaps incredibly) the state of Lebanon still exists, but it remains more strongly rooted in confessionalism than ever, and each new conflict seems to increase the number of religious and political minorities within its boundaries.

Not only are there significant Arab minorities in religious terms, but also there are many peoples living in Arab countries who do not speak Arabic. The conquering Islamic armies succeeded in converting the

Kurdish peoples of the northern Levant and southern Anatolia and the Berber peoples of the mountains of the Maghreb to their faith, but these peoples maintained their own languages. In virtually every Arab state today there remain residual cultural or religious minorities who were never converted to Islam and who have never been fully assimilated into what might be termed 'mainstream Arab culture'. There are also many immigrant minorities, such as the Armenians, who have entered the Arab countries and who have been allowed to live there within geographically distinctive cultural enclaves.

It is also important to note that the Islamic faith spread to nations beyond the Arab realm and that twentieth-century states such as Turkey and Iran, although Islamic by faith, are not Arab countries by either language or culture. From a Western perspective both Saudi Arabia and Iran are states whose societies are steeped in the traditions of Islam and whose rulers throughout the 1980s have been fervent in sustaining traditional Islamic values. However, as noted above only one of them is an Arab state and, as will become clear below, the characteristics of the Islamic faith which they guard are very different.

Islam and the Arabic language may be central features of many Arab societies, but from an Arab perspective language and religion are as often seen as the source of cleavages within society as they are perceived to be integrating forces. Only in the face of certain external threats from the West have they become bonding forces between Arabs.

From an Arab perspective the peoples of their region are perceived as strongly differentiated from one another in terms of their beliefs and social norms. The divisions of Islam within the Arab world are strong and of long standing. For example, one only needs to read the accounts of the travels of the fourteenth-century Arab geographer, Ibn Battuta, to realize that feuding and distrust between Sunni and Shia Moslems is long standing and that rivalry between different Islamic sects owes as much to inherited traditions as it does to doctrinal differences. The original schism between Sunni and Shi'ite groups dates from a dispute over the political succession to the Prophet. The details of the dispute are less important than the outcome which was a division of the Islamic faith into a tolerant orthodox Sunni group (the word comes from *Sunnah* or 'well trodden path') and a more secretive and less tolerant Shi'ite group, whose most well known historical cult was the Assassins of the eleventh to the thirteenth century (responsible for murdering hundreds of Sunni religious leaders in their bid for power).

Sunnism is dominant in North Africa, the Western Levant and Saudi

Arabia, while Shi'ite sects are in the majority in eastern Iraq, Yemen and southern Lebanon. Shi'ism is of course the dominant doctrine of major Islamic nations such as Iran. Following the Islamic revolution in Iran, their religious leader, the Ayatollah Khomeini, resurrected an extremist Shi'ite faction known as *Hizbollah*, whose radical doctrines and actions have been accused by several of the Sunni Arab states as being responsible for Islamic uprisings. A good example of the factionalism within the Islamic community is the way that *Hizbollah* acted in West Beirut in 1988, when they defeated both the Sunni Muslim militias and the pre-existing Shia army known as *Amal*. This illustrates well the internal tensions and struggles for power which exist within Islamic society, in one particular geographical context. The issue of fragmentation and lack of unity amongst the peoples of the Arab region is taken up once again in Chapter 3, where it is examined in terms of the consequences it has had for the region's development.

Perceptions of Arab politics and government

Islam also serves as a reference point in the description of the progressiveness of Arab governments. Until the late 1970s, Western commentators were liable to characterize Arab states as lying somewhere on a modern/traditional continuum, on which those states which were most traditional would also be seen as clinging most strongly to the Islamic faith. This led to the unfortunate tendency for the West to equate Islam with what might be termed 'conservative' or 'anti-modern' political systems. In Western terms the Islamic faith has therefore often been perceived as being a potential hindrance to the 'modernization' of the Arab world. The 1980s showed that in practice radical Islamic political movements were not against progress and change, but against the Western vision of what constituted development. From an Arab perspective Islamic fundamentalism has therefore been an influence which has not only distinguished their region from the West, but has also proved to be a divisive force within the Arab world.

Muslims are required to give allegiance to the state only if the state is judged to be faithful to the holy books of Islam. Consequently, in many Arab states where political opposition in the form of conventional political parties has been difficult, it has been to Islam that those disillusioned with their lot have often turned. Equally it has been through Islamic religious leaders that political dissatisfaction has often been articulated against those Arab leaders who have taken a line too close to Western positions. The result has been that Arab governments have ignored Islamic teachings at their peril. For example, it was an Islamic Fundamentalist group which was responsible for the assassination of President Sadat of Egypt in 1981, claiming that his policies were contrary to fundamentalist teaching. His successor, President Mubarak, while not imposing Islamic law, was quick to claim that Egyptian law was 90 per cent Islamic (thus not alienating the Christian Copts), and that the *Sharia* (the holy rule book of Islam) was the basis of the Egyptian constitution.

A particular tension which impinges on development in the Arab world is that which exists between the interests of the state (whose modern powers and limits were largely imposed by Western nations at the end of the colonial era) and the interests of Islam, interpreted by fundamentalists as a world system of thought and action which long predates the current territorial divisions of the Arab world. There is therefore much potential for conflict between the ruling political elite within each of the Arab states and the forces of Islamic fundamentalists. The ruling elite inevitably seek to legitimize their position in terms of the 'success' of the state in economic, social or spiritual terms. By contrast Islamic fundamentalists tend to seek pan-Arab and pan-Islamic objectives, and often interpret state problems as a reflection of bowing to Western influences and failing to enforce Islamic law.

Unlike other developing nations, the Arab countries faced additional problems in the early years following independence in asserting their new identities (Ajami, 1982). New political leaders not only had to be seen to have thrown off the economic and social heritage of their colonial predecessors, but they also had to prove an independence from Western philosophies of development and nationalism. Their legitimacy depended on their ability to interpret political independence both in national terms and in a manner compatible with pan-Islamic ideals. It is this which, amongst other factors, has fuelled a desire by so many Arab leaders to prove that they are leaders not only of individual states but of the Arab world.

Of course the way in which Islamic law has been interpreted by politicians has varied greatly from one country to another. Consider for example the differences which exist between Saudi Arabia and Libya. Saudi by any standards is a conservative and strict Islamic monarchy, which prides itself on guarding the holy cities of Islam, forbids women to drive cars and even demands separate bus stops for men and women, but nevertheless maintains a foreign policy which, like its national economic linkages, is distinctly pro-Western. Libya, another desert oil state, is by contrast ruled by a revolutionary army commander, Colonel Gaddafi, who has imposed Islamic law and interpreted the *Sharia* in such a way as to justify the nationalization of Western assets as part of a drive to establish an Islamic state free of all symbols of Western culture.

One cannot conclude any appraisal of the differences between Western and Arab perspectives without specific mention of the Israeli–Palestinian conflict. This is referred to on many occasions throughout this text (and in most detail in Chapter 3), because it has been the source of continuing conflict in the region throughout the latter half of the twentieth century. It has been an issue which has united the Arabs on most occasions behind the stateless Palestinian people, and which has frequently engendered financial and military intervention by the West, chiefly in support of the Israelis. Conditioned by Western familiarity with Judaeo-Christian traditions and spurred by Western guilt over the holocaust and the mis-treatment of the Jews by West Europeans over many centuries, the West has been largely supportive of the intrusive presence of a Zionist state in territory that was previously part of the Ottoman region and later the British mandate of Palestine. By contrast one of the most powerful forces uniting the Arabs of the twentieth century has been the call to defend Arab lands from Western-supported Zionist settlers. Arab military forces from many different states have been willing at one time or another to fight against the Israeli army. Considerable Arab finance has been given by the oil-rich states to support the Arab frontline states with Israel and, prior to 1990, any sign of wavering from a pro-Palestinian position brought condemnation from the Arab states. Egypt, for example, in being willing to sign a peace treaty with Israel in 1978 was forced to leave the Arab League and was ostracized by the other Arab states for most of the 1980s. To the Israelis and to many Western politicians the forces seeking to establish a Palestinian state, such as the Palestinian Liberation Organization, were branded as terrorists, while to the Arabs they were perceived as both freedom fighters and the legitimate political representatives of the Palestinian people. This matter is considered in more detail in Chapter 3.

Arabs have good reason to perceive Western involvement in the Arab world as much more than the post-colonial legacy which is how a Western presence might be interpreted in other developing countries. It is also an alien cultural influence which not only is sourced in a religious belief system quite other than Islam, but also is supportive of a political and religious minority group which has taken away Arab land and rights. From a Western perspective, up until 1990, the Arabs were often labelled as anti-Zionist and anti-Western, without there being a full appreciation of the historical basis for their position. For some Western observers, Arab efforts to establish a Palestinian state and to reduce, if not to remove, the powers of the Israeli state have been seen as a tragic obsession which has distracted Arab energies and resources from other major development issues. To many Arabs the struggle remains a central symbolic fight which must be won to achieve liberation from the continuing cultural, economic and military presence of the very forces responsible for the region's under-development.

The Iraqi invasion of Kuwait and the subsequent war altered the position of some Arab states on the Israeli–Palestinian conflict. The Palestinian leadership, by taking the side of Saddam Hussein in the war for reasons which cannot be explained here, alienated not only some of their Western support but also the Saudis and Kuwaitis. Support from Saudi Arabia for Jordan was cut and after the war most of the Palestinian community living in Kuwait were forced to leave. While the President of the United States claimed in the aftermath of the war that a new window of opportunity had been opened on the Palestinian–Israeli dispute, 1991 ushered in a new set of political alliances in the Arab world and brought to an end four decades in which the Palestinians' fate served as one of the few unifying political forces within the Arab world. It remains to be seen whether the limited 'self-determination' offered by Israel's Yitzak Rabin in the Gaza enclave and the Jericho district, after the prolonged 1992–3 Peace Accord talks, will in fact be allowed to grow into the reality of an independent Palestinian state.

Conclusion

The objective of this book as stated earlier is to consider the development achievements, problems and potential of the Arab world. This chapter has shown how difficult it is to undertake such an assessment as a Western observer. Arab development cannot be meaningfully assessed until resources are defined, not in terms of Western standards and mechanisms for capital accumulation, but in terms of Arab perceptions and objectives. This is not to suggest that all Arab economic or social policy has been wise or is justified (sadly this seems seldom to have been the case), but it is to suggest that patterns of Arab resource use have not always occurred in ignorance of the alternative paths which

Western economies might have followed. Development has to be assessed in terms of the self-defined objectives of the people involved if analysis is to be more than an evaluation of Western progress in incorporating the resources and peoples of other regions into a Western-dominated world economy.

Key ideas

1 Western and Arab perceptions of what constitutes 'development' vary greatly.
2 Comparison of economic and social indicators for different states reveals considerable inequalities within the Arab world.
3 Oil is perceived by the West as the Arab world's greatest natural resource, but not all Arab countries enjoy oil wealth, and oil has often been a dividing influence between Arabs.
4 Islam is perceived by the West as a common cultural influence on the Arab world, but Arabs are more aware of the divisions within their lands between ethnic groups and religious minorities.
5 The interests of Islam and of the Arab state are often hard to reconcile.

2
The colonial legacy

Introduction

The nineteenth and twentieth centuries have been the scene of many dramatic and fundamental changes to the social and economic structure of the Arab world. From an Arab perspective many of these so-called developments were neither desired nor desirable forms of change. This was because they were externally imposed and served only to further the development of foreign, and usually Western, interests. Ironically Arab countries have often found that the struggle to resist Western political and cultural advances has in itself required internal social, economic and political re-organization and the adoption of concepts alien to traditional Arab ways. Thus the Arabs have found themselves faced with the dilemma of either directly accepting or adapting Western 'modernizing' ideas or officially rejecting them only to find that the process of opposition to Western models of development in itself has acted as an indirect catalyst to change in Arab society.

The Ottoman influence

Independent development of the region by the Arabs themselves ceased in many senses in the sixteenth century. By then the vast majority of the Arabs had become subjects of the Ottoman empire. The Ottoman empire had expanded consistently from the mid-fourteenth century, from a core area which lay in what is now thought of as Turkey, to a position by the mid-seventeenth century in which it controlled much of

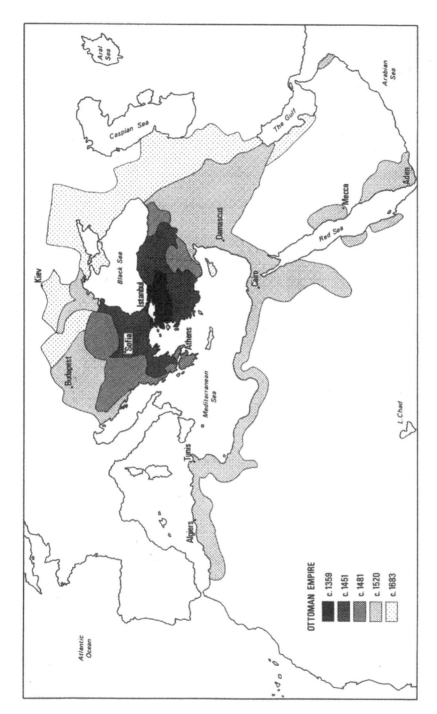

Figure 2.1 Growth of the Ottoman empire

the Mediterranean and Arab world. Only after the Ottomans had already conquered the Balkans and parts of southwest Asia such as Azerbaijan did they turn south. In order to ward off the Portuguese maritime colonizers who were threatening Muslim pilgrimage and trade routes they incorporated the territories of Syria, Egypt and the Hijaz in 1516 and 1517. The Arabs of North Africa were at the same time facing increasing problems from the Spanish and called on the Ottoman navy to help them. Once the Ottoman admirals had repelled the Spaniards, they divided the eastern Maghreb into three of their own regencies (Algiers, Tunis and Tripoli) (Figure 2.1).

Although the Ottomans endorsed Islam as the dominant religion of their empire and enforced varying forms of *Sharia* law throughout their territories, they were aware of the 'multi-national' character of their empire. Including as it did Turks, Kurds, Arabs, Slavs and Greeks it also spanned different religious and linguistic groups, and chose in general to favour tolerant co-existence of these groups through permitting minorities some freedoms within what was known as *millets* or communal organizations. For the Arabs the system meant that neither their traditions and distinctions from the Ottoman rulers were erased nor those of the minority groups which lived in Arab lands. That the Ottomans looked west as much as they did east is evident from the way in which they recruited their military elite in the sixteenth century. These were by tradition young men taken exclusively from Christian families, converted to Islam, and then trained in Istanbul for a particular branch of military or administrative service. This recruitment system seems to have worked effectively for a time and was favoured for many reasons, not least of which was the fact that it ensured that the chief administrators and key troops were intensely loyal to the sultan. It did have the effect, however, that people born as Muslims, and consequently the majority of Arabs, were virtually excluded from the most senior administrative and military positions in the Ottoman empire, and as a result had no experience in high government. This and other aspects of the Ottoman system therefore meant that their empire was Islamic, but certainly not Arab.

Although the Arabs in no way became slaves of the Ottomans, the very fact that they were Ottoman subjects (that is to say, they were incorporated within a vast centralized political system which was not Arab) had far-reaching consequences for their subsequent economic and political development. It would appear that for most of four centuries the Arabs accepted that, with the passage of the ultimate military power in the Islamic world into the hands of the Ottomans, so also had passed the process of ultimate decision-making and government on issues relating to their position within the world. At the same time, however, the existence of limited local autonomy within the individual Ottoman provinces such as Egypt or Syria led to the establishment of limited territorial power groupings, some of which were subsequently to form the basis for 'Arab' nationalist independence movements.

European influences

If Ottoman rule produced the conditions for the political under-development of the Arabs, it was the increasing European influence on the region which was to have a far more profound effect on the economies and societies of the Arabs in the nineteenth and twentieth centuries. As one Arab analyst has remarked, 'The Ottomans may have been our rulers, but at least they did not take our lands.' The Ottoman military presence may have delayed the European invasion of the Arab world. Equally it is possible that the expanding European empires of the eighteenth and early nineteenth centuries did not perceive the Arab lands as containing great riches or raw materials in such abundance as could be found in other more easily colonized parts of the globe. Rivalry between the European powers was another force delaying incorporation of the Arab territories. For example the French invasion of Egypt in 1798 was reversed three years later by the British and the Turks. Settler colonization was only attempted on any scale by the French, Italians and Spanish in the Maghreb. Algeria was invaded by the French as early as 1830, but other parts of the Maghreb were not finally dominated by any one of the European powers until late in the nineteenth century or early in the twentieth century. Morocco was not finally partitioned between the French and the Spanish until 1912, the same year as Italy occupied Libya.

Elsewhere the rise of European influence did not seem initially to demand any fundamental changes in Arab society or economy. Britain's interest in much of the region at the beginning of the nineteenth century seemed largely to relate to its concern to maintain its trade routes to India. The leaders and peoples of the different provinces and minorities within the Ottoman empire found themselves courted by the European powers who were preparing for the overthrow of the Ottomans. To semi-autonomous regions of the empire, such as Egypt, most European technical and social innovations were therefore seen as offering the means to emulate European successes in repelling the Ottomans and to achieve similar military and industrial might as existed in the West.

Egypt is a useful example of a country which was eager to repel both Ottoman and European colonial advances, but which did so largely by seeking a development path which followed European lines. Mohammed Ali, the Egyptian ruler during most of the first half of the nineteenth century (1805–48), bought European machines, encouraged the growth of new Egyptian enterprises and sought to create the circumstances for a small-scale Egyptian industrial revolution. Mohammed Ali also made considerable efforts to reform the agriculture of the country. Over a million acres of land were brought under the plough during his reign and he was responsible for major irrigation works, like the great Nile Barrage just north of Cairo. He also established elements of a new education system introducing aspects of European knowledge and the teaching of technical subjects, mathematics and the Italian language. He insisted that this educational training was essential for access to key administrative and military posts. These and other measures helped Egypt grow in economic strength in a truly remarkable fashion.

In parallel to his other efforts, Mohammed Ali purchased European weapons, created a conscript army trained by foreign instructors and increased Egypt's military might. Early in the nineteenth century the Egyptians successfully repelled a British expeditionary force and later became the dominant military influence in Syria and Palestine. In the long run, however, Egypt's new prosperity and its key location as guardian of the Suez canal led to increased European intervention culminating in the British occupation of the country in 1882. British interest in the region was not, however, like that in other parts of Africa and Asia. To quote the words of Lord Curzon, a British Foreign Secretary of the early twentieth century: 'The Arabs should be our first brown dominion and not our last colony.'

French colonization of the Maghreb

While most of the Arab lands continued to lie within the Ottoman empire and did not come under Western rule until the collapse of the empire after the end of the First World War, the Arab countries of North West Africa did not manage to maintain their independence for as long from colonial powers. Where submission to Western military or economic might occurred, it soon led to the geographical restructuring of these Arab economies, initiating the construction of new transport systems and administrative centres. These were located in such a way as to service an export economy, designed to facilitate the trading strategies of European colonial powers. The spatial development of Tunisia is a good example of this process (see Case study B), and one which has parallels in the experiences of Morocco, Algeria and to a lesser extent Libya. In Morocco and Tunisia the pre-colonial settlement systems were clearly oriented around Islamic power centres such as Fes and Kairouan. These Islamic cities were located in the interior of their respective countries, but within a few decades of colonization this pattern had been displaced by colonial settlements in the coastal areas close to the main European-built ports.

Case study B

'Modernization' in Tunisia – the colonial experience

Throughout the nineteenth century the French looked on Tunisia as a natural extension to their Algerian territories (annexed in 1830). Using the pretext of frontier clashes, and in the face of financial and administrative disorder in Tunisia, the French occupied the country and made it a Protectorate in 1881. They vigorously encouraged agricultural colonization by French settlers to substantiate their right to govern the country (in the face of Italian opposition). It was this policy which first caused them to instigate major changes in the spatial structure of the country.

Transport networks are of great importance in channelling and redistributing a nation's human and physical resources in a manner perceived as appropriate by those governing a country. The Tunisian railway system is one example (Figure 2.2) of the close association which can be observed between the development of a transport system and the evolution of colonial interests. By 1888 the French had established a line between Tunis and Bone (Annaba) in eastern Algeria. This was followed by several branch lines to the most promising agricultural areas in northern Tunisia such as the lower Medjerda valley and the Cap Bon peninsula. A second phase of construction came in the early twentieth century following the discovery of valuable mineral deposits (phosphates,

Case study B (*continued*)

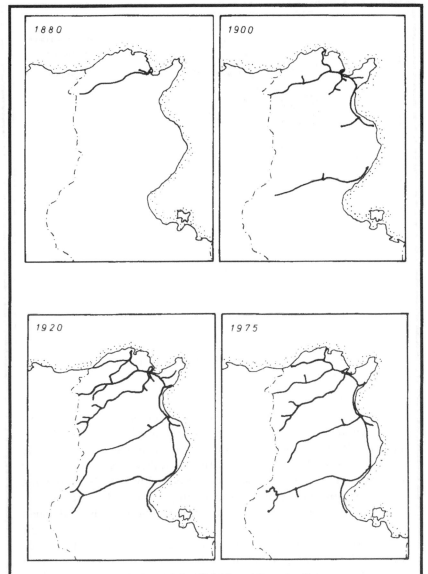

Figure 2.2 The evolution of the Tunisian rail network
Source: Findlay (1980)

Case study B (*continued*)

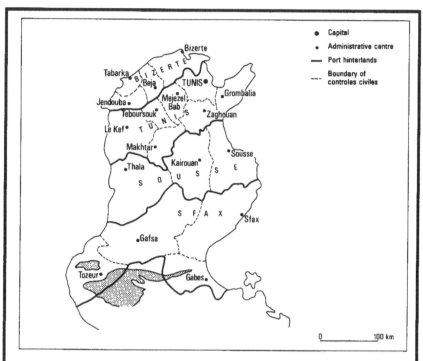

Figure 2.3 Port hinterlands in the Tunisian Protectorate
Source: Findlay (1980)

iron ore, lead and zinc) in the interior of Tunisia. Lines were
rapidly constructed to link these interior deposits with the nearest
ports where these minerals could be exported to Europe. It was
not until 1911 that the major cities of Sousse and Sfax were directly
linked by rail. Even at its zenith the railway system only had
one main north–south track compared with four lines making
east–west links between the four main ports and the interior.

In order to govern their new territories, the French established
a hierarchy of administrative centres at key points along their
transport networks. Some centres were designed largely to service
the rural colonial population; others were involved mainly in
enforcing colonial law and order among the Tunisian tribes,
thus securing the uninterrupted exploitation and transportation

Case study B (*continued*)

of mineral deposits. A symbiotic relationship therefore arose between the French-built transport system and a number of key points in the settlement hierarchy which were selected to organize and control the movement of physical resources along the vital lines of communication. The four port hinterlands coincided with the macro-administrative units or 'controles civiles' (Figure 2.3). Within each 'controle' one settlement was selected to be an administrative headquarters. They also became the urban centres in which the French introduced Western institutions such as colonial law courts, hospitals with Western medicine, Western style schools, libraries and post offices. The centres naturally soon began to experience rapid population growth as people from surrounding areas migrated to these places of greater opportunity.

In the northwest and northeast of Tunisia, where settler colonization was most intensive, a finer territorial division was established to serve the interests of the expatriate population. It was here too that innovations such as banks and Western schools were first introduced, thus creating a spatial imbalance in the timing and intensity of social and economic change, with the north being systematically advantaged relative to the south. This bias has long outlived the colonial era, and will be hard to reverse because of the difficulty of restructuring fundamental features such as the urban and transport structures established by the French. It is also important to consider critically the long-term consequences of the 'modernization' process. It established an economy oriented to meet the needs of France rather than those of the indigenous Arab population. The effects of the apparently 'beneficial' introduction of health and education services included the undermining of many aspects of Tunisian rural society through the imparting of urban skills, the more rapid growth of population than of the rural economy, and the precipitation of rural–urban migration to the coastal cities favoured by the colonial regime.

Consider the case of Morocco. The traditional Islamic capitals of Morocco are Fes (founded in the ninth century) and Marrakech which was established by the Sanhaji nomads in the eleventh century. From these cities of the interior, the Moroccans in the fifteenth, sixteenth and seventeenth centuries resisted repeated invasions from Iberia. The Portuguese controlled a few coastal ports such as Ceuta, Tangiers and Safi, but Arab power remained intact in the interior of the country. The authority of Moroccan sultans waxed and waned over the centuries, depending on their abilities to overcome tribal factionalism. In the eighteenth and early nineteenth centuries the sultans were increasingly forced to use diplomatic rather than military means to avoid European colonization. Their main tactic was to play off one European power against another. The last great sultan, Moulay al-Hassan (1873–94), reduced tribal dissidence in the face of the strong threats of external colonization but was unable to stop European economic intervention. After his death tribal rebellions gave the French the opportunity to land troops to 'protect' their citizens and property. Continued economic penetration and tribal disorder led to a growing French military presence in those areas with the greatest agricultural and mineral potential. Long after economic links had been established between the two countries, Morocco was finally declared a French Protectorate in 1912.

French colonialism meant that French settlers came to own over a million hectares of the best agricultural land such as in the Fes-Meknes plain, the Gharb, the Chaouia and the Souss plain. The massive land transfer which colonization brought about, as well as a failure to invest and encourage development in areas of traditional rain-fed agriculture, led to the impoverishment of a large part of the Moroccan rural population. It has been estimated that by 1931 one-third of all Moroccans were landless: large numbers of them left the countryside in search of employment in the growing colonial cities of Casablanca and Rabat. Colonization was therefore one force which contributed to the uneven pattern of land holding (see Chapter 6 for more details) which was to become such a severe handicap to the subsequent agricultural development of the Arab world.

A second major motive for colonization by the French was exploitation of Morocco's considerable phosphate deposits. Phosphate reserves were nationalized by the French, thus excluding the possibility of private capital interests gaining ground. Instead the French colonial administration gained an important source of income and developed phosphate exports as a major earner of foreign currency. Little effort was made to develop other industries. Samir Amin, a well respected commentator on the economies of the Maghreb, notes that by the end of the French colonial period no proper basic industry existed (Amin, 1970). The colonial legacy in Morocco was therefore an economy whose agricultural sector had been traumatically disrupted by the effects of settler colonisation, and an undeveloped manufacturing sector, suppressed in the interests of sustaining a market for French industrial products in Morocco by minimizing competition from indigenous industries.

This thumbnail sketch of the economic under-development which Morocco experienced as a result of French colonization could be repeated in a similar form for Algeria or Tunisia. The spatial consequences of establishing an economically extroverted system was to change the demographic and settlement balance of all three countries. In Morocco this is evident from considering the relative fortunes of the old cities of the interior with the much more rapid growth of coastal settlements. Table 2.1 shows that all urban areas grew rapidly during the first half of the twentieth century as rural–urban migration redistributed population towards the major cities. The rate of growth was much greater in the coastal cities, however, where colonization had brought increased prosperity, than in the traditional Islamic capitals of the interior. The pattern evident in Table 2.1 is symptomatic of the much more fundamental redistribution of wealth and opportunities which occurred as a result of the colonial era. Colonization meant a spatial redistribution in favour of the coastal zone, and the creation of an infrastructure which was to establish self-enforcing economic processes which would produce a pattern of spatially uneven development.

Table 2.1 Populations of coastal and interior cities of Morocco at the beginning and end of the colonial era (in thousands)

	c.1910	1936	1960	Growth 1910–60 (%)
Interior cities				
Fes	100	144	216	+116
Marrakech	70	190	243	+247
Meknes	40	75	176	+340
Coastal cities				
Casablanca	12	247	965	+8,282
Rabat-Sale	44	115	303	+589
Tangiers	20	–	142	+610
Kenitra	–	18	87	–

Source: adapted from Abu Lughod (1980, 153, 248)

In some senses submission to Western colonial power brought with it developments which were beneficial to the Arab peoples concerned. In the Maghreb the application of Western medical knowledge helped to

reduce the worst effects of the diseases which had afflicted the population. Epidemics of cholera and typhoid became less common and
medical knowledge was used to tackle tuberculosis and trachoma. The
benefits of Western medicine should not be over-emphasized, however.
For example, in Algeria amongst the French colonial population it was
not until three decades after the colony was established that death rates
fell to below the birth rate. For the Arab populations the benefits of
Western medicine took much longer to produce a sustained change in
life opportunities. Demographers such as Rouissi (1977) suggest that in
the Maghreb as a whole it was not until the 1890s that one could detect
a sustained decline in the death rates of the Arab population. This was
partly due to the geographical unevenness of medical provision, but also
because of the low standard of living of large parts of the population.

Another change agent associated with the colonial era was the
diffusion of Western-style education (in contrast to the long established
koranic schools or *medersa*). Again it is possible to interpret this as both
a positive and a negative influence. Western schooling for Arab children
may have accelerated rural–urban migration through the implicit urban
bias of European educational values and skills. Most analysts, however,
would agree that reduction of illiteracy and the increase in numeracy
were key steps to the formative development of the peoples of the Arab
world. Investment in the education of their population has certainly
been one of the development policies which independent Arab states
have done well to pursue in the latter part of the twentieth century and
in the face of other demands on their resources.

Despite the gains listed above, the net effects of the colonialization
were negative. Even after colonial rule had come to an end, the
economic and geographical structures of the colonial era often remained
largely in place. Thus, long after their departure, foreign powers
left behind spatial structures and processes which at best hindered
development and more often caused under-development.

The partitioning of the Middle East

For a variety of reasons settler colonization was never contemplated
on any scale in the eastern part of the Arab world. As has been noted
above, the dominant European powers (France, Britain, Russia and
Germany) were all acutely aware of the weakness of the Ottoman
empire in the latter part of the nineteenth century, but moves by any
one of them to annexe a major part of the Arab territories of the empire

sparked off intense rivalry amongst the others, which usually resulted in blocking movements. Nevertheless, certain spheres of influence were established in relation to their respective national interests. One method of achieving limited intervention was to claim the right to protect minority interest groups. Thus, for example, Russia claimed the right to protect all Orthodox Christians throughout the empire; France took a special interest in the plight of the Maronite Christian communities of Lebanon and, following massacres of the Maronites in 1860, landed French troops in Beirut to 'keep the peace' and to establish a Christian governor in Lebanon (see again Case study A); and some British politicians developed a sympathy for the Zionist cause to establish a Jewish homeland in Palestine.

During the First World War the European powers increased their efforts to ensure that they were each well placed to benefit from the collapse of the Ottoman empire. The details of the intrigues of the Western powers have been reported in detail in other texts and are not repeated here. To understand the subsequent economic development of the Arab states it is, nevertheless, necessary to understand a little of how the state boundaries of the Arab region came into existence, since these boundaries and the states which they enclose have been the source of ongoing disputes in the region. These current-day disputes, inherited from the period of colonial military intervention, have absorbed a massive amount of Arab attention and resources. These resources might otherwise have been capable of contributing much more positively to the welfare and development of the Arab economies. This matter is discussed further in Chapter 3.

The French and British made a series of conflicting promises to the peoples of the region during the First World War. The most significant of these were

1 the McMahon correspondence of 1915 and 1916 which promised Arab independence in return for a revolt against the Ottomans by the tribes of the Hijaz,
2 the Balfour Declaration of 1917 which confirmed British support for the idea of a Jewish homeland in Palestine, and
3 the Sykes–Picot treaty of 1916 which was an agreement between Britain, France and Russia about how they would partition the Middle East.

Clearly there were elements of these promises which could not be reconciled one with another, as well as innumerable contradictions and assumptions on the part of the parties making the promises. They arrogated the right to determine the future of the Arab lands without wider consultation with all the parties involved. The result was a period of uncertainty after the First World War when various groups tried to claim the promises which had been made. Finally Britain and France (with the international approval of the Treaty of Lausanne) imposed more permanent frontiers on those territories whose external relations they largely controlled.

In the five years between 1918 and the Treaty of Lausanne, some political units emerged which indicate what the political structure and subsequent economic development of the region might have been like had it not been for Western intervention. In 1916 the hereditary rulers of Mecca, the Hashemites, started the Arab revolt and were influential in speeding the defeat of the Ottomans in Palestine and Syria. In 1918, in line with the McMahon correspondence, they established Damascus as their newly independent capital. Then with the support of nationalists from Palestine and Syria, they demanded an independent Arab state which in principle might have stretched over large parts of the Western Levant. To the north, minority populations, which were neither Arab nor culturally close to the rising tide of Turkish nationalism, briefly succeeded in wringing from the defeated Ottomans, under the Treaty of Sevres, the right to small, separate Kurdish and Armenian states. An Armenian Independent Republic existed between 1918 and 1920, before being conquered and absorbed into the Soviet Union. In the southern corner of the Arabian peninsula, the collapse of the Ottoman empire created a political vacuum in territories of little concern to any of the European powers. Here the fiercely individualistic mountain tribes of North Yemen regained their independence within a traditional *Imamate*. The state had no centralized authority, and by remaining cut off and largely isolated from the outside world survived for over forty years as little more than a loose confederation of tribes, united only by a weak theocratic leadership. The strategic significance of some of the ports on the southwestern coasts of the peninsula meant that here British interests were strong. As a result British control of Aden and the Trucial coast cut off the interior tribes of Arabia from their seaboard on the Indian Ocean.

The political developments described above are interesting because they all reflect a change in political organization from the Ottoman period, with a shift towards smaller units, each developing much greater cultural cohesion in linguistic and religious terms. Undoubtedly many

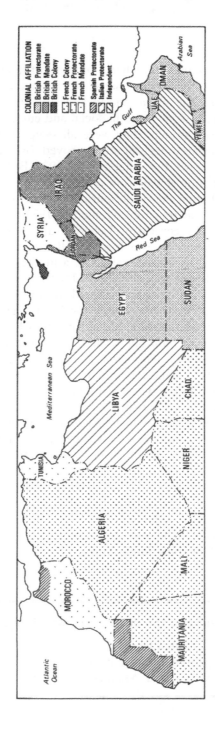

Figure 2.4 European colonization and Arab independence

of these proto-states would have had immense problems in making economic progress, in the Western sense of the concept, given their limited resource bases and often land-locked situations. Their boundaries would have made some sense in social and cultural terms. With the exception of North Yemen, they were all swept away by various pieces of Western-inspired legislation approved by the League of Nations. Figure 2.4 shows the outcome. In 1922 Britain was given mandated powers to control Palestine (which because of the Balfour treaty was carved off from the neighbouring terrain to the east of the Jordan). Transjordan and Iraq were created in 1921 and 1926 respectively. France was given control of Lebanon and Syria. Faisal, the Hashemite ruler in Damascus, was forced to flee in the face of bitter nationalist opposition. His brother raised an army against the French, but was offered the internal governorship of Jordan in return for peace. Faisal, who had fled from Syria, was offered the throne of the British mandated territories of Iraq. Meanwhile the French greatly enlarged the territory of Lebanon to include, along with the Christian Maronite population, large areas of both Shi'ite and Sunni Muslim populations as well as other minorities. This created circumstances justifying a strong French presence to make such a heterogeneous state governable.

One major exception to the pattern described above is the Kingdom of Saudi Arabia. This great desert territory was not perceived either to contain major physical resources or to have strategic military importance. In addition, the emergence of a strong and astute Arab leader who established control over the territories without presenting a threat to the West, and who was able to point to British promises of independence for the provinces of the Hijaz after the Arab revolt against the Ottomans, meant that here, unlike elsewhere in the Arab world, a powerful and independent state did emerge. Case study C describes and analyses the forces that produced and shaped the character of twentieth-century Saudi Arabia.

Case study C

The House of Saud and the emergence of a kingdom

The western part of Arabia has always been of particular significance to the Arab peoples because it includes the holy sites of the Islamic faith, in Mecca and Medina. Arabs throughout their

Case study C (*continued*)

history have been willing to make immense sacrifices to protect
these holy places from external control and influence. Even the
presence of some 500,000 foreign troops on Arabian soil during
the Iraq–Kuwait war was very strictly monitored. As one Western
envoy remarked, this was probably the first war in history without
war brides.

In the early twentieth century the Wahabbis, followers of an
eighteenth-century reformer whose puritanical beliefs had led him
to take a forceful stand against all influences which he perceived
to be non-Islamic, re-asserted their influence in the Najd region
of Arabia. Nomadic peoples were encouraged to abandon non-
Islamic practices, and in adhering once more to the strictest
interpretations of the koranic law (*Sharia*), were settled in self-
sufficient military agricultural communities. Abd al-Aziz Ibn Saud

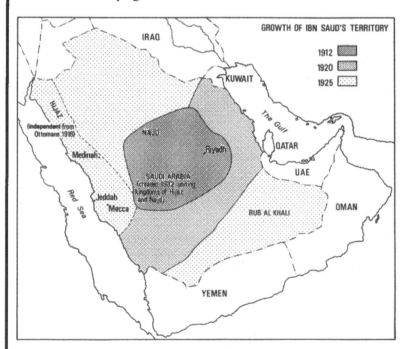

Figure 2.5 The emergence of the Saudi state

Case study C (*continued*)

(1880–1953) harnessed the Wahabbi willingness to die fighting to convert other Arabs to their own standards, and extended the area of Wahabbi influence to the Hijaz. In re-imposing fundamentalist Islamic principles and practices on Mecca and Medina, the leader of the House of Saud also successfully established himself as the undisputed leader of the desert kingdom which was to become known as Saudi Arabia.

The kingdom was perceived by the Western powers of the time as having only minor economic and strategic significance. As a result no external power intervened to halt the ambitions of the House of Saud. The main threat was therefore internal, and arose from the difficulty of reconciling the development goals of what initially was a relatively poor state with the puritanical beliefs of the religious movement which had brought the Saudi family to power. Religious objections were found to motor cars and telephones and there were also pressures from some Wahabbis to wage a religious war against the British in Iraq and Transjordan. Eager not to earn the potentially disastrous animosity of a Western military power, which was already sympathetic to the overtures of the former rulers of the Hijaz, and not wishing to be cut off from the 'benefits' of Western technology, the Saudis were forced to go to war against some of the very forces which had helped them to establish their authority. Abd al-Aziz Ibn Saud therefore succeeded in creating a kingdom where the *Sharia* became the framework for the law, but where royal decrees could intervene in specific cases to permit certain types of development which he considered desirable. As a result of the production of oil from 1938 onwards, this delicate balance became even more difficult to sustain, since it strengthened the forces favouring the adoption of Western innovations. In practice the real and perceived wealth which the Saudi royal family were to receive greatly enhanced their power and position as rulers of a strictly Islamic state and guardians of Islam's most holy places.

The political and economic geography of Saudi Arabia is therefore very different from that of nearly every other part of the Arab world. This is so for many reasons, but perhaps the key difference is that Saudi Arabia is a state which emerged, not as a

Case study C (*continued*)

political tool imposed by Western powers to exploit strategic or physical resources, but as a result of internal social and cultural processes which culminated in a 'modern' state (note how differently the word 'modern' is used here by comparison with Case study B). The House of Saud was equipped with the machinery to direct the minds and resources of the peoples in relation to Arab aspirations and objectives. This is not to say that all Arabs would agree with the development path which Saudi Arabia has followed, but it is to suggest that the spatial structure of the Saudi economy reflects the decisions of certain Arabs as to how the country's physical and human resources should be used. The superficial results of this are evident in the comparison of key locational features of the country. In contrast with the colonies of the Maghreb, Saudi Arabia maintained greater strength in the interior of the Kingdom in terms of its urban structure and transport network.

Unfortunately the history of Saudi Arabia is the exception rather than the rule in the Arab world. Most contemporary Arab states owe their boundaries, the definition of their state power, and many of their problems as political units, to the Western powers which partitioned most of the map of the Arab world after the First World War. A glance at the shapes of the states shown in Figure 2.2 shows that many of the boundaries of the Middle East are straight lines, reflecting military or external political judgements about the partitioning of territories and little regard for the cultural mosaic of the lands concerned. Examples include the boundary between Egypt and Sudan, Libya and Tunisia, Syria and Iraq, and Kuwait and Iraq. The result has been the creation of many state boundaries which divide former national groupings such as the Kurds, while at the same time defining political territories which include very heterogeneous populations with no sense or possibility of coherence between state and national identity.

The internal government of the mandated territories was largely left to indigenous Arab leaders, while France and Britain determined their international policies. With the exception of Palestine, there was very little Western settlement in the mandated territories of the Machrek, or Arab east. In some senses the Western influence on the states of the

Arab east, might therefore be thought of as being very limited by contrast with the indelible imprints of settler colonization which had been imposed on the economies of the Maghreb. At the geo-political level the legacy for subsequent economic and social development of Western influence in partitioning much of the region could not have been greater. The influence of Western powers in the military and political sphere in enforcing a permanent territorial division of the Arab world in spatial units, largely alien to the indigenous development needs of the Arab population, was immense in shaping the framework of future Arab development. The Western geo-political units which were implanted were the ones within which state nationalism would subsequently emerge. They were the context within which internal power struggles would appear and consequently they set the scale on which many of the region's political movements would evolve. Writing some seventy years after this set of political units was created, perhaps the most surprising feature is that it has largely endured. The power of the state, once fully established, appears to have been very great, with even the most poorly integrated and incoherent of units, such as Lebanon, surviving decades of strife between the warring ethnic and religious units.

After the Second World War most of the mandated territories were at last to achieve their full independence from the colonial powers. Although there have been many proposals by the leaders of these states to abolish the boundaries imposed by the West and to form some pan-Arab coalitions within which new political and economic developments might emerge, these proposals have seldom led to any real changes. A champion of these types of proposal was President Nasser of Egypt who, in his self-appointed role as political leader of the Arab world, announced the emergence of a United Arab Republic involving first of all Egypt and Syria (1958) and later Egypt and North Yemen. Later, Colonel Gaddafi of Libya sought the marriage of his country with Egypt (1972–3), then Libya and Tunisia, and most recently, Libya and Morocco (1984). After decades of proposals and counter proposals North and South Yemen declared themselves to have re-united in 1990.

None of these diplomatic attempts to dissolve boundaries imposed on Arab lands by the West have proved durable, except for the merger of the two Yemens. Clearly some of the merger proposals were unrealistic from the outset and were little more than irredentist gestures. Some were genuine, however, and had some economic logic as well as a social and political basis. The failure of plans to abolish the boundaries imposed by the colonial era has been due in most cases to the underlying strength of specific interest groups within the original states. Long after the colonial era it would appear that the policy of 'divide and rule' has remained in place with those forces sustaining separatist state power proving stronger than the political and economic forces favouring re-unification and integration. These failures have proved particularly frustrating for the Arab world in the 1990s, in view of the apparent success of the two Germanies in negotiating re-unification after more than four decades of separate development.

Conclusion

It would not be too strong to conclude that the legacy of the colonial era in the Maghreb has been a spatial and economic infrastructure which has made development of these states extremely difficult without the continued orientation of these lands towards the West. In the Arab east the legacy of Western intervention in the latter part of the nineteenth century and the early part of the twentieth century has been continued internal political instability as a result of the way that states were constituted. It has also created the potential for ongoing conflict rather than co-operation between rival Arab states, vying for access to divided physical resources and strategic positions. In addition the insertion of a non-Arab state, Israel, in what was the British mandated territories of Palestine, has produced conditions of continuing instability and the ever-present threat of war throughout the whole region. This long-running dispute, discussed in more detail in the next chapter, has like so many other conflicts in the Arab world resulted in the tragic diversion of scarce physical and human resources.

Almost half a century after the official independence of most Arab lands, the continued fragmentation of the Arab world into the state units initially imposed by the West seems unlikely to be reversed in any substantial way. The Arab experience of the West has left a legacy of internal conflict and external distrust. The development implications have not only been economic extroversion, but also widespread political instability.

Key ideas

1 The French colonization of North Africa produced patterns of dependence on the French economy which the Arab states have found hard to reverse since independence.

2 The basis on which the colonial powers partitioned the Middle East served their interests well, but created many inherently unstable political territories for the post-colonial era.
3 Attempts to abolish colonial boundaries have largely failed, although the concept of recreating one large Arab nation remains a potent one which appeals to many Arabs.

3
Political constraints to economic development

Introduction

The previous chapter emphasized the way in which the colonial era created structures (spatial, social, economic and political) which on balance militated against the smooth economic development of the Arab region in the second half of the twentieth century. It is important to recognize, however, that many forces internal to the Arab world have also slowed the pace of economic advance. At times they have substantially set back what, in Western terms, would seem to be indicators of economic well-being. This chapter considers three such forces, and attempts to show how interaction between these 'weaknesses' and external political and economic interests may have seriously diverted Arab human and physical resources away from an agenda for balanced economic development. The three forces to be considered are as follows:

1 the continued religious and ethnic divisions of the peoples living in Arab lands, which have sustained economic fragmentation of the region as a whole, with divisive influence on states within the region and even on communities residing in the same city;
2 the numerous conflicts which have arisen over the last four decades which have diverted scarce resources into military structures and have often resulted in the devastation of the economic and physical infrastructure, as well as imposing grievous loss of life and great human suffering;

3 poor government, lacking in large part any democratic authority and
   hence tending to impose control through the centralization of power
   and failing critically to question development strategies in ways which
   might have led to policies being fruitfully adapted to match local
   needs and opportunities.

In considering these topics it should be remembered that the reason
for discussing them is their bearing on the economic development of
the Arab world. Many other political, cultural and moral issues are
associated with the topics included in this chapter, but cannot be treated
in detail here, as they lie outside the main theme of this book.

Religious and ethnic divisions

In the introductory chapter the division of Islam into Shia and Sunni
sects was discussed. Figure 3.1 offers a more detailed description of the
fragmentation of Islam.

The theological and historical basis of these religious groups is less
important to the theme of the present chapter than two other characteristics. First, the way in which the Islamic world has traditionally dealt
with fundamental differences within the faith has been through the
distancing of opposing views, as represented by the concentration, and
sometimes segregation, of religious minorities in particular places or
localities. Prior to the arrival of the Western concept of the 'nation state'
these minorities often controlled many of their own affairs, enacting
within their community or locality their own religious practices. This did
not necessarily mean economic isolation from surrounding peoples – for
example the Kharijite peoples (Figure 3.1) of the island of Djerba were
an important trading community enjoying extensive commercial links.
Under the Ottoman empire the spatial separation of religious minorities
was one means by which a decentralized political system was able to
allocate responsibilities in a hierarchical fashion. This allowed considerable autonomy to minorities to organize their internal affairs.

The establishment of the state as the dominant organizing force
intervening between the world economy and the local community
changed the power balance within Arab society. Suddenly the divisive
effect of religious differences took on a new significance since, in the
absence of government by democratic consensus, sectional interests of
the dominant group (Sunni Muslims in most of the Arab world) became
more important because they tended to control the received state

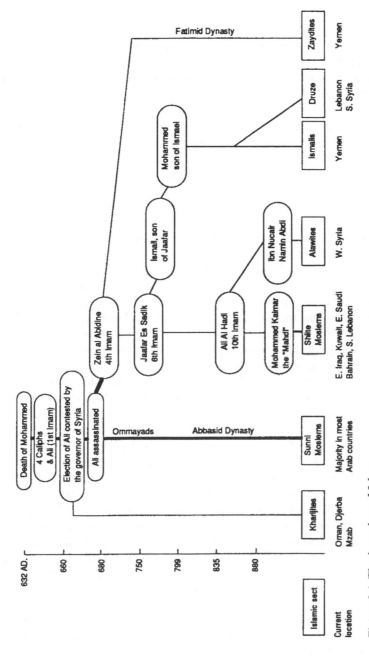

Figure 3.1 The branches of Islam
Source: adapted from Boustani and Fargues (1990, 28)

apparatus, with all the power that this implied to direct economic development. This could be done to the benefit of some sections of society more than others, through the ability of state government to determine, amongst other things, how productive economic processes such as agriculture and industry should be organized and to whose advantage.

The unequal (and often unjust) distribution of state power between different religious groups implied the inevitable emergence of coercive influences and conflicts between religious minorities and those in power. The spatial concentration of many religious minorities has served as a convenient factor, on the one hand facilitating the actions of those in power in maintaining control over minorities, while on the other hand making more likely regional contests of power within the state. The most striking example of this in the early 1990s has been the struggle between the Shia populations of the marshlands of Iraq and the Sunni dominated regime in Baghdad. In Syria a more unusual situation exists where the political leaders are Alawite (Figure 3.1) drawn from the western provinces of the country. In Syria only 11 per cent of the population belong to this sect. Radical Sunnis have repeatedly called for a holy war against the Alawite leadership, with the most violent uprising being the 1982 revolt in Hama. The revolt was crushed by the ruling Alawites using great force.

The second feature of religious division in the Arab world which needs to be considered is the special relationship which exists between Islam and political leadership. Followers of Islam are encouraged to see political and spiritual leadership as one and the same thing, as long as those in control of the state are judged to be faithful to the laws laid down in the holy books of the religion. The incorporation of groups with significant religious differences in a system of states was therefore problematic from the outset, as it required either a departure from the Islamic view of a fusion between spiritual and political leadership or a constant undermining of state authority by minority groups looking to community religious leaders as the key figures of political reference. This, combined with the spatial dimension discussed above, effectively encouraged the creation of states within states of the kind that has proven so disastrous in Lebanon since 1975 (see once again Case study A).

Many of the points made above with regard to divisions within Islam apply equally to other minorities not following the Islamic faith or as defined on ethnic grounds. The arrival of the state as an organizing apparatus provoked conflicts between those in power and marginalized ethnic minorities such as the Kurds, Berbers, Nubians and Armenians. Under the dominant influence of Arab leaders the states in which these groups reside have seen their languages undermined by state education systems privileging Arabic over minority languages and often forcefully repressing attempts within minorities to foster their cultural identity. In many states massive energy and investment has been sunk in their efforts to govern minorities, which prior to political 'independence' had little to do with their Arab neighbours. Consider for example the case of Sudan where the Arabs of the north, although making up the largest single group in the country and effectively dominating all aspects of the country's government since independence in 1956, still only accounted for 9.5 million of Sudan's population of 24.2 million in 1989. For the first twenty-seven years of its existence the state of Sudan was engaged in a bitter civil war between the Muslim north and the fragmented Christian and pagan minorites of the south. If the resources used to sustain the civil war had been available to invest in agriculture, Sudan might well have been able to escape some of the worst effects of the famine in the drought years of the 1980s.

The cost of conflict

The fragmentation of the Arab world along religious and ethnic grounds has served as a major force distracting governments from the task of economic development and necessitating the diversion of enormous human effort and scarce financial resources into the task of maintaining control of the state apparatus. In the absence of a democratic consensus this has meant the use of force and coercion and the development of military machinery to police the state.

Table 3.1 attempts to measure the extent to which those in control of the Arab states have felt it necessary to allocate funds from their state budgets for military purposes. The effect of the Gulf conflict of 1990 and 1991 was to heighten the contrast, already evident in the table, between the states of the Maghreb with their moderately low military budgets and the high military expenditures of the other Arab states. Table 3.1 shows that only Tunisia and Morocco spent more on education than on defence. To put it another way, only these two countries invested more on human development than on potentially destructive projects.

Table 3.1 State budget and military expenditure in selected states of the Arab world in the 1980s

Country	State budget as percentage of GNP[a]	Percentage of state budget devoted to		
		defence	education	health
Egypt	47	21	11	2
Iraq	59	50	n.d	n.d
Jordan	48	28	12	4
Kuwait	52	13	11	6
Morocco	31	16	17	3
Oman	46	47	8	4
Saudi Arabia	54	36	13	5
Syria	44	55	11	1
Tunisia	39	10	15	7
United Arab Emirates	35	45	9	7
Yemen Arab Republic	22	22	16	4

Source: derived from Boustani and Fargues (1990)
Notes: [a] Average for the years 1981–8.
         n.d., no data.

A feature not evident in Table 3.1 is the extraordinary difference between the Arab countries and other developing regions of the world. In the 1980s the Arab states spent approximately 13 per cent of their GNP on the military. The next highest military budget was in Eastern Asia at 5 per cent of GNP, while the states of sub-Saharan Africa devoted only 2 per cent of their GNP to the military. Clearly military expenditure represents a much greater foregone opportunity in development terms for the Arab states than for most of the rest of the developing world, and this assessment does not even begin to count the cost in lives and ruined infrastructure of the use of military force.

Internal conflicts in the Arab states have accounted for only a fraction, in budgetary terms, of the cost of conflicts between states. The patterns of expenditure in Table 3.1 to a large extent map the history of warfare between the states of the Middle East since 1945. Iraq and Iran waged an immensely costly eight-year war (1980–8) which resulted in hundreds of thousands of Iraqi deaths and absorbed on average 28 per cent of the country's gross domestic product (GDP). Saudi Arabia and the other Gulf states were concerned that the Iran–Iraq war would spread and invested heavily in arms in the 1980s, ironically only to deploy them in 1991 in the Iraq–Kuwait conflict. In the aftermath of the war arms sales to Saudi Arabia and the other Gulf states reached new heights.

Many of the conflicts in which the Arab states have been involved have been orchestrated or aggravated by the intervention of external military forces. The scale of this external involvement is perhaps best indicated by the following statistics: the United States has spent more on military assistance to Israel and Egypt since 1945 than on any other states except for Vietnam. Between 1945 and 1990 it has been estimated that the United States gave Israel US$29.5 billion in military aid. Egypt, the next largest beneficiary, received US$13.3 billion over the same period. During the cold war between the superpowers, the Soviet Union and China were also very active in the region reflecting the perceived strategic significance of the area. Conflicts between Middle Eastern states in the region have therefore tended to be viewed by the superpowers as being of global importance. Their involvement in most of the conflicts has served to heighten the destructive force of the region's wars. Sadly under President Bush's so-called New World Order it was again the Arab countries which were first to experience the military consequences of being involved in a conflict involving the world's one remaining superpower.

It has been estimated that the consequence of the allied bombing and the United Nations sanctions on trade with Iraq had the effect, in development terms, of reducing the standard of living of ordinary Iraqis from one similar to the most prosperous parts of Eastern Europe to that of India. In particular, damage to electricity related infrastructure was peculiarly serious for Iraq, since the flatness of the country has meant that water and sewage in some areas does not flow by gravity but requires pump assistance. Thus even isolated villages were affected by the destructive bombing campaign, since damage to the electricity system stopped pumps distributing water to melon fields and to live-stock. Even two years after the war, animal products were still too expensive for most Iraqis to buy, and the collapse in living standards was inevitably followed by an increase in child stunting and in infant mortality rates.

The 1991 Iraq–Kuwait war had longer-term implications for develop-ment right across the Arab world. This was because it represented the most serious split to occur between Arab states since the Second World War. Any semblance of Arab unity which might have existed before the war was destroyed, and bonds of economic interdependence between many of the Arab states were seriously damaged. This was evident, for example, in the repatriation of migrant workers from the oil states to those countries sympathetic to Iraq's position, most notably the expulsion of Palestinians from Kuwait and of an estimated 1 million Yemenis from Saudi Arabia. In order to pay for its war costs, and for other reasons, Saudi Arabia moved away from supporting oil prices within the OPEC oil cartel (discussed in more detail in Chapter 4) and started producing much larger quantities of oil at lower world prices, while patterns of inter-Arab aid flows changed to the benefit of Egypt and the disadvantage of Jordan.

Palestine and Israel

One conflict which has often united the Arab world and which has also had profound development implications for the region has been the ongoing struggle between the Israelis and the Palestinians. Surrounding Arab states have been drawn into the conflict in wars in 1948, 1956, 1967 and 1973. The Israeli invasion of Lebanon in 1982 was also directly related to their dispute with the Palestinians.

The British mandated terrritory of Palestine was the only region of the Arab world which did not become part of an independent Arab state in the years following the Second World War. The Western nations, still shocked by the atrocities against the Jews during the Nazi holocaust, were even more sympathetic than before the war to the concept of forming a separate Jewish homeland in Palestine. This idea was in line with the ambitions of the world Zionist movement and the Balfour Declaration of 1917 (see Case study D), but ran contrary to the spirit of the Sykes–Picot treaty of 1916 which had agreed that an integrated mandated territory should be set up rather than two separate states, one for Jews and one for Arabs. It was perhaps predictable that the West would favour plans supporting two states, since the boundaries of Western Europe had themselves been drawn on the basis of the concept of the 'nation state' – that is, a national concentration of people sharing a strong cultural heritage within a political territory over which they exert the rights of self-determination. The problem in Palestine was that the territorial unit was claimed with varying degrees of legitimacy by two separate peoples – the Jews and the Palestinian Arabs. To make matters more complex the vast majority of Jews did not live within the territory which they sought as their homeland. The Balfour Declaration was not unique since other national groupings in the Middle East, such as the Kurds and Armenians, were also promised support at one time or other to form independent states. Unlike the Jews they never achieved all the necessary conditions to realize or sustain this goal, and consequently found themselves absorbed into new states founded on a multi-ethnic (pluralistic) basis.

Case study D

The Balfour Declaration

> The Foreign Office
> London
> November 2, 1917

Dear Lord Rothschild,

I have much pleasure in conveying to you, on behalf of His Majesty's Government, the following declaration of sympathy with Jewish Zionist aspirations which has been submitted to, and approved by, the Cabinet.

'His Majesty's Government view with favour the establishment in Palestine of a national home for the Jewish people, and will use their best endeavours to facilitate the achievement of this object, it being clearly understood that nothing shall be done which may prejudice the civil and religious rights of existing non-Jewish communities in Palestine, or the rights and political status enjoyed by Jews in any other country.'

I should be grateful if you could bring this declaration to the knowledge of the Zionist Federation.

Yours sincerely,

Arthur James Balfour

In practice it was the forces favouring partition and the separate develoment of Arab and Jewish communities which were to prove by far the stronger after 1945. In 1947 a United Nations plan was disclosed which sought to establish two states in what had been Palestine. The plan was heavily affected by the distribution of Jewish settlement which had taken place during the first half of the twentieth century under the guidance of the Jewish National Fund. This fund had systematically bought land blocks in Palestine in order to facilitate the establishment of colonial villages populated by Jews immigrating to Palestine – initially from Eastern Europe but later from Germany and other countries where anti-semitic movements were active.

Two important effects followed the planting of the new Jewish rural settlements (known as *moshavim* and *kibbutzim*). First the process served to absorb and disperse the ever growing Jewish population. This grew mainly by immigration from a mere 5 per cent of the population of Palestine in 1880, to 22 per cent by 1940, and 31 per cent by 1948. Second, the moshavim and kibbutzim established an identifiable structure to Jewish settlement, making it possible for the Jews to lay claim to a significant consolidated territory which could be shown to 'belong' to Jews as a basis of claiming a state of their own. The efficacy of this approach became evident in 1947. Although the Jews remained less than a third of the population of Palestine they were allocated about 55 per cent of the land under the United Nations plan.

The indigenous Arab population and surrounding Arab states, not surprisingly, did not accept the United Nations' proposals. When the British withdrew a war between the Israelis and Arabs ensued immediately. The result was victory for the newly formed Jewish state and the expansion of its territory to an area some 20 per cent greater than had been suggested by the United Nations. Many of the Arab peoples fled from the Israeli occupied areas, while most of the remainder of what the United Nations plan had indicated would become an independent Arab state (in the highlands to the north and south of Jerusalem – known to some as the Samarian and Judean highlands) was occupied by Jordanian troops and administered by Jordan for nineteen years. It is important to note that it was not just Israel but also surrounding Arab states which at this time precluded the creation of a separate Palestinian state.

Thus began a process of Arab Palestinian refugee movements that was to be repeated following the 1967 'Six Day War', the 1971 war in Jordan, the 1982 invasion of Palestinian controlled areas of Lebanon and the 1992 'liberation' of Kuwait by the anti-Saddam coalition. The initial Palestinian exodus from the newly established Israeli state was to some extent balanced by waves of Jewish immigration from other Arab countries and from Europe. Between 1948 and 1956 some 850,000 Jews migrated from other parts of the world to Israel. In the 1960s and

1970s the immigration flows continued, but came primarily from other parts of Asia and Africa. Almost half a million Jews migrated to Israel between 1967 and 1990. In the late 1980s a new flood of Jewish immigration began from the Soviet Union/Commonwealth of Independent States. Between 1989 and the end of 1991 no less than 350,000 Soviet Jews (*olim*) arrived to settle in Israel. Encouraged to migrate by Israel, it has been estimated that up to 1 million *olim* could settle in Israel by the mid-1990s.

The 'development' of the Israeli state during its first forty-five years has therefore been associated with massive population movements, both of Palestinian refugees and of Jewish immigrants. This dual mechanism is only one of many aspects of the Israeli–Palestinian conflict, but it serves well, in the context of the theme of this book, to illustrate the consequences of a policy of separate development. This has meant in practice the continued 'development' of a Zionist state to the benefit of Jews over Arabs, in what was the mandated territory of Palestine. That the goals of Zionism were not satisfied by the events of 1948 has been made evident from the ongoing expansion of the Israeli state.

The Six Day War of 1967 resulted from a pre-emptive strike by Israel against her neighbours in an atmosphere in which Israel perceived the threat of a joint attack by Egypt, Jordan and Syria. During the war Israel conquered the whole of the West Bank territories, the Sinai peninsula, as well as the Golan Heights. The occupation of these territories created new waves of refugees. In the case of the West Bank some of the refugees were second-time movers of persons already displaced by the 1948 war.

Once again the policy of the Israeli government appeared to be the occupation of the land that had been gained, in order to claim that once it had been settled it was rightfully part of the Israeli state. As a consequence new settlements were planned and installed in the West Bank, the Golan and Sinai. Gaza and the Samarian Highlands were perceived as Palestinian enclaves, but in the West Jordan valley and in the Golan, where refugee departures had been considerable, the pace of the early (1967–77) settlement programme was greatest. Settlements in this early phase were located primarily in relation to the Allon Plan which sought to establish a 'secure' border along the Jordan river. A small number of agricultural villages were established to lay extensive territorial claims to the West Jordan valley, while leaving the rest of the West Bank relatively untouched. By 1976 only just over 3,000 Jewish settlers had moved to these villages. The dominant thrust of Israeli resettlement in the Occupied Territories was in East Jerusalem which the Israelis had officially annexed in 1967, hence seeking to impose on this area Israeli civilian rather than military law. The colonization of East Jerusalem is a topic not discussed in detail in this book, but the scale of Israeli settlement (around 70,000 persons) shows the level of commitment to this symbolically significant space, and the intention to integrate it in a permanent fashion into the Israeli state. It is interesting to note that unlike East Jerusalem the rest of the West Bank has remained, even in Israeli terms, an Occupied Territory.

The election of a right wing Likud government in 1977 brought sweeping changes to the West Bank settlement policy. Some analysts such as Rowley (1984) have argued that even without a change of government a major extension of the settlement policy would have occurred and that land sequestration in the agricultural core of the West Bank was already planned or under way. Between 1977 and 1981 widespread settlement throughout the West Bank was promoted in line with the Drobles Plan. Whatever the origins of the plan, the implantation of about fifty settlements in the densely settled highland areas heightened tensions with the Palestinian population, since considerable land areas were now being taken over not just for agricultural villages but for dormitory commuter settlements (known as *Yshuv Kehillati*) to serve Jerusalem and the cities to the west. Much of the settlement programme was co-ordinated by the religious group Gush Emunim. In addition to establishing villages and suburban settlements, the Drobles Plan sought to introduce six new towns at strategic locations across the West Bank. In these towns 60 per cent of the families were to be employed in industry, handicrafts and tourism, while the remainder would work in services or become commuters.

The pace of settlement accelerated during the second Likud government (1981–4), but then slowed again in the late 1980s when a coalition government was formed which included those who opposed further settlements. The existing settlements continued to be strengthened and by 1990 about 70,000 Jewish settlers had moved to the West Bank. This still represented only a tiny fraction of the total population of the West Bank, but in terms of territorial control the 118 new settlements represented a significant extension of the Israeli state, which in many respects was far more threatening to the Palestinian population than the Israeli military presence.

Two external factors also slowed the pace of the Israeli West Bank settlement programme. The first was the reduced numbers of immigrants arriving in the mid-1980s, at the very time when some of the younger Israelis were becoming disillusioned with living in a Zionist inspired state and were themselves seeking to re-emigrate. Without a flow of new immigrants, and in the face of very rapid growth of the Arab population due to the high levels of natural increase amongst the Palestinians, the settlement programme lost momentum for a time. In addition the cost of the programme proved an increasing burden for the fragile Israeli economy, with its very high levels of inflation (often running at over 100 per cent per annum) and its heavy dependence on foreign loans and aid. By the end of 1991 it was estimated that the settlement programme had cost Israel US$3 billion. The re-emergence of a dominantly right wing government in the early 1990s, combined with a new influx of immigrants from the Soviet Union/Commonwealth of Independent States, led to a brief but forceful new phase of settlement building in 1991 and 1992. This was halted neither by the freeze on loan guarantees from the United States nor by the establishment in 1991 of a new round of Arab–Israeli diplomatic negotiations. These were initiated in the aftermath of the war with Iraq, seeking to bring about a peaceful and more long-lasting resolution to the Israeli–Palestinian conflict. Only in 1992, following the re-election of a Labour-led coalition government in Israel, did the building of new settlements come more decisively to an end.

An understanding of the nature of the conflict between Israel's policy of encouraging Jewish immigration and settlement in the Occupied Territories and the perception of the West Bank amongst Palestinians is important for identifying the reasons for the continued existence of two quite separate development strategies. These are held by two nations for territory which they 'share'. To Israel the significance of the West Bank is primarily strategic. Although attachment to the area arises out of historical and religious factors, the original settlement of the West Jordan valley was to secure a new frontier. Subsequent settlement of other parts of the West Bank has been justified by some on the basis that the highland terrritories of Judea and Samaria would form a military threat to the security of nearby cities such as Tel Aviv and Jerusalem, should they be controlled by an unfriendly neighbouring state. This argument has been somewhat undermined by the events of 1991 when Iraqi missiles showed that Israeli cities could be targeted from much further afield.

To the Palestinians, the West Bank remains the core of their perceived homeland, but the land itself has significance as a major resource enabling the remaining population to support themselves from agricultural production and the region's water resources. These assets are particularly important in the absence of any real investment in the development of industry in the West Bank in the four and a half decades since 1948. Removal of land for the building of Israeli settlements has therefore denied the remaining Palestinians access to what they perceive to be one of their few remaining sources of earning a semi-independent livelihood. According to Coon (1990), in 1967 nearly 90 per cent of the land of the West Bank was privately owned by Palestinians, while by the end of 1991 about 60 per cent of the land had fallen into the hands of the Israeli state.

In the absence of a means for strategic investment by the Palestinians in the West Bank, the economic development of the region to the benefit of the remaining Palestinian population has been constantly frustrated. As a result the other main resource of the Palestinian people – its human resources – has been dissipated over the years. With few job opportunities in the West Bank many Palestinian men in the early 1980s sought work instead inside Israel. Israel with its low rates of natural increase was only too glad to tap this reserve of cheap labour to work mainly for low wages in blue collar jobs in Israel's cities. Since the Palestinians continued to live in the West Bank, this meant the establishment of a complex daily commuting pattern of over 100,000 people from the West Bank into Israel. Because of this process the Palestinian workers have sometimes been referred to as 'nomad labour' (Portugali, 1989). The more significant aspect of this development was that it represented a breaking down to some extent of the pre-1967 boundary, and it justified in part the building of new road networks integrating the West Bank with Israel.

'Nomad labour' was only one way in which the 'under-development' of the West Bank's human resources took place. More significant in numerical terms has been the exodus of refugees and economic migrants. A map of the distribution of the Palestinian population in the 1990s would show that the majority of Palestinians no longer live within the boundaries of the former mandated territory of Palestine but are spread across the Arab world as refugees and migrant workers. The latest phase in their diaspora came in 1991 when, following the Iraqi defeat in Kuwait, most of the 400,000 Palestinian community were forced to flee to Jordan.

Jordan has become the single largest concentration of Palestinians. By 1992 many of Jordan's Palestinian population had never lived in

Palestine, consisting instead mainly of persons born to refugees. Many had lived their entire lives as refugees within the camps set up in 1948 and 1967. Case study E describes the largest of these camps at Baqaa, some 15 miles to the north of Amman.

Case study E

Baqaa refugee camp

The green and white bus bumped uncomfortably along the road north from Amman to Suweilah and the Baqaa camp. The bus was very busy, crowded with women and children from the camp. Their conversations were more than casual chat – they were in miniature the story of the camp itself.

Hassan got his diploma. He leaves next week for Qatar. He got a job in the airport at Doha. I don't expect we will see him for a long time. He will stay with Mohammed Nasir when he arrives.

It won't be long until Mohammed comes back for his wedding. Only Ahmed can come home for it. The construction company won't let Bassam have time off. Anyway he hasn't enough money yet to come back. After all with the baby coming they need all they can get.

Did your Ibrahim get a job yet? Abdelaziz said that there were some jobs coming up in the hospital in Kuwait.

From another part of the bus I heard some other voices.

Little Jamila is so ill. A chest infection. It is so damp in these houses.

It seems to be more muddy than ever in the camp this year. Mohammed wanted us to go to Kuwait. He says there are nice flats there. But if his job ended or if something went wrong where would we go then? At least we have the little place here. Perhaps some day, Inshalla (God willing) . . .

Case study E (*continued*)

I seemed to hear that 'Inshalla' all day in the camp.

Suddenly the bus broke the crest of the ridge and I could see below me the curious sight of the vast Baqaa refugee camp. It had always struck me as incongruous the way in which this sprawl of make-shift homes for some 60,000 refugees sits totally dwarfed by two large white radar dishes. It was more than twenty-five years since the so-called 'camp' (far from the temporary feature which the term implies) was established, yet the sophisticated electronics of the dishes continues to remind the refugees daily of the threat of further conflict.

I descended from the bus, relieved to see the familiar and welcoming faces of the Palestinian students who had agreed to undertake a sample household survey on my behalf. Unlike the paved road from Amman, inside the camp the roads were mud tracks. The January rains had transformed the usually dusty lanes of the camp into a quagmire. A stone, undoubtedly intended as a missile, landed in a puddle not far off, and Mohammed, one of my student friends, turned to rebuke the child responsible, before leading me away rapidly to another part of the camp.

Plate 3.1 Housing in Baqaa refugee camp, Jordan

Case study E (*continued*)

Plate 3.2 Main street of Baqaa refugee camp

I was taken to Mohammed's house. Like most Palestinian homes in the camp its walls were made of mud, permanent building materials being officially forbidden within the camp. In the courtyard of the house, as I sipped a glass of sweet tea, Mohammed asked 'How many brothers do you think I have?' I raised my head to look around. 'No. No. They are not all here,' he said. 'Many work in Kuwait. I have fourteen brothers, but my father has married twice.' He smiled and turned to discussing the survey.

The survey had gone well. Most households had agreed to fill in the questionnaire on emigration to the oil states and on levels of return migration. The quantitative results of the questionnaire were what I had come for, but somehow the people on the bus had already told me what to expect. Nearly all households in the camp had a large number of their young men absent, many of them as migrant workers in Kuwait or Saudi Arabia. Those who had returned to the camps (and other parts of Amman) were

Case study E (*continued*)

> mainly older men who had been born on the West Bank or in
> Palestine before 1948. But the families of the refugees had mainly
> been born in Jordan. They were refugees, but they had never lived
> in Palestine. Baqaa was their place of birth, but not their home-
> land. One day they hoped to return to Palestine.
>
> *Source*: details of the survey described in this case study are published in
> A. Findlay and M. Samha (1986) Return migration and urban change. In
> King, R. (ed.) *Return Migration and Regional Economic Problems*, 171–
> 84, London: Croom Helm

Approximately one-third of Palestinians continue to live in refugee camps representing their failure and that of the world community to resolve their dispute with Israel satisfactorily. They continue to lack an independent homeland, the right to a passport and the abililty to follow a development path chosen by their own leaders. Many external commentators doubt the viability of the West Bank as a suitable territory for an independent state, even were the Israelis to permit the region limited autonomy. By 1992, although the Labour-led Israeli government had agreed to talk to their Arab neighbours and the Palestinians over the future of the Occupied Territories, the West Bank had become increasingly integrated into the Israeli state as a result of the new physical infrastructure developed during the years of occupation. Also, many of the most capable young Palestinians had already left, either as refugees or economic migrants. Those advocating the West Bank as a territory suitable for an 'independent' Palestinian state have to contend with the fact that the territory is landlocked, has no major mineral or energy resources, has virtually no industrial base, has an agricultural system which is heavily labour intensive and is heavily dependent on access to the proximate and lucrative Israeli market. Thus the West Bank economy remains very vulnerable to fluctuations in the volume of migrant remittances and external aid received.

In the context of the immense development problems faced by the Palestinians of the West Bank, it is not surprising that despair has often given rise to violent actions described in the West as terrorism but seen by many Palestinians to be no more than part of their fight for political freedom. The words of Abu Jihad, the co-founder of the Palestinian resistance organization Al Fateh, express the rhetorical position of many Palestinians:

> Of course we dream of peace. I want to be with my family in peace. But our dreams are destroyed by reality. We are obliged to fight and to remind the world that we exist. If we were to wait for the world's conscience, we would never return.
>
> (Abu Jihad, quoted by Dimbelby, 1979, 239)

The merger of Al Fateh and several other Palestinian organizations such as the PFLP in 1964 to create the Palestinian Liberation Organization (PLO) under the charismatic leadership of Yasser Arafat gave hope for a time to the populations of the Occupied Territories that either a Palestinian-led military machine or guerilla actions could progress their cause. The dismemberment by Israeli troops of the PLOs self-created state within a state in Lebanon dashed this venture and was to stimulate at least two other responses. One was the search by the PLO for a diplomatic solution through attempts to engage support for the Palestinian cause amongst Western nations: the other was the igniting of the *intifadah* in the West Bank.

The *intifadah* (Arabic for 'uprising') commenced in December 1987 and quickly became a widely supported uprising throughout the West Bank. Ironically this revolution of stones was to give the PLO far greater legitimacy than any previous 'violent' acts, since it demonstrated to the world the continued suppression of the peoples of the West Bank by the Israeli army. It marked a coming of age of a Palestinian generation no longer willing to have Israeli development plans imposed upon them (Usher, 1991). Some five years later the *intifadah* continues to make government of the West Bank by the Israelis very difficult. It has led to the interruption and later cessation of 'nomad' labour movements of Palestinians to work in the Israeli economy. Indeed it has been argued by the Israeli geographer, Newman, that the *intifadah* has been the single most important force in aiding the Palestinian cause throughout the world.

The perceived transformation of the PLO from a terrorist organisation to one representing a national struggle for independence, the declaration of the independent Palestinian state at the Algiers summit of Arab leaders in November 1988, and the decision of the United

States to hold informal meetings with representatives of the PLO
have been spurred by the popular uprising.

(Newman, 1991, 49)

The *intifadah*, or its equivalent, remains a necessary but not sufficient
condition for the creation of a Palestinian state. In the meantime, in its
first five years, the *intifadah* has been associated with under 100 Israeli
deaths and over 1,000 Palestinian fatalities. It has also re-established in
the eyes of the world the 1967 boundary of the Occupied Territories.

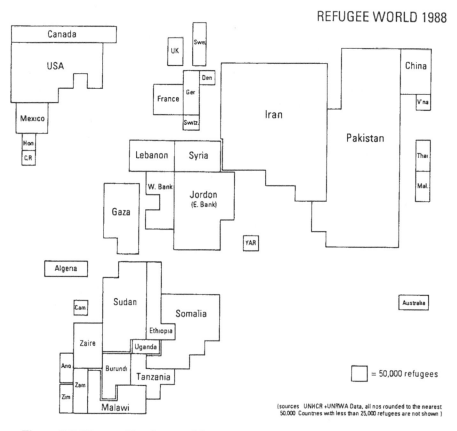

Figure 3.2 The world refugee crisis
Sources: UNHCR and UNRWA data
Note: All numbers rounded to the nearest 50,000. Countries with less than 25,000
    refugees are not shown.

Refugees

As stressed at the outset of the previous section, the Palestinian–Israeli dispute is only one of many conflicts which has burdened the Arab world and which has distracted its leaders from single-mindedly seeking the economic development of the region. Of course physical security remains a prerequisite for sustained economic development. Many governments in the region have found that, even if they have not been involved in direct military action, their economic efforts have been frustrated by the need to accommodate large populations of refugees fleeing from conflicts in other countries. Indeed it is easy to substantiate the claim that Islamic countries, more than any others, have had to bear the cost of providing aid to the world's refugees. The basis of this claim is clear from Figure 3.2. It can be seen that, contrary to what one might think from listening to media reports of Western politicians discussing the influx of refugees to Western Europe, by far the largest concentrations of refugees are to be found in the Islamic realm. Iran and Pakistan for over a decade hosted the world's largest refugee population – about 5 million Afghans. Following the Iraqi invasion of Kuwait, Jordan temporarily and at considerable expense hosted major refugee populations (not technically recognized as refugees by the world community because they were Asian and Arab migrant workers fleeing Kuwait) and, after the war, Iran and Turkey received refugees fleeing from the Kurdish and Shia areas of Iraq.

Weak government

Given the immense difficulties facing Arab governments in terms of the problems listed above, it is perhaps dubious to use the term 'weak' government to describe one of the main problems responsible for the slow rate of progress towards economic development in the Arab region. Many of the reasons for 'weakness' have arisen from the inherited agenda of problems discussed in the last chapter and the first part of this one. Before proceeding to consider some of the development problems which have stemmed from weak government, it is also important to stress that the term 'weak' does not mean that these governments have not used force. As illustrated earlier in this chapter, military force has all too often been the preferred means of implementing unpopular government policies in the Arab region. Arab governments have, however, been weak, first in the sense of lacking legitimacy to implement their development policies (few being able to claim any popular mandate to govern in the democratic sense of the word), and second in the sense of having very limited economic flexibility to choose which policies to implement.

In 1992 only Egypt, Tunisia and Jordan could claim to have a multi-party political system which was based in some way on 'free' elections (and there are many who would challenge the representativeness of the elections even in these states). In most of these countries the power of elected politicians (and in particular of opposition groups) remains rather limited, but at least the voice of a political opposition is allowed to be heard. At its most successful in the Arab world, democratic movements have been allowed to flourish on the condition that opposition parties may contest elections but not win them. For example, in Egypt's three general elections held since the 1980s the electoral system has meant that the ruling National Democratic Party has retained a minimum of 85 per cent of the seats in the People's Assembly. In addition the Egyptian constitution states that presidential decrees have the power of law, hence investing vast powers in the president and reducing the power of the elected assembly.

Some Arab countries have tried and failed to sustain multi-party political systems. Kuwait, for example, experimented with a limited degree of democracy prior to suspending its elected assembly in 1986. It took the Iraqi invasion and post-liberation pressures for reform to get the emir of Kuwait to hold fresh elections for October 1992. Algeria held free multi-party elections early in 1992, but when the Islamic Party gained a majority of seats the army re-took control of the country and the democratic experiment was suspended, at least temporarily. In other states such as Libya, Iraq and Mauritania military regimes continue to command absolute power, defying the trend in Eastern Europe for popular uprisings to overthrow totalitarian regimes. The leaders of other countries such as Saudi Arabia make little apology for having avoided experiments in democracy. In the wake of the Kuwait–Iraq conflict the voices demanding reform have grown louder in Saudi Arabia, leading to the proposal for a consultative assembly. This assembly will not initiate legislation but only inform King Fahd of its views. King Fahd, announcing the proposal to launch the assembly, explained his position thus in March 1992:

The democratic system that is predominant in the world is not a suitable system for the peoples of our region. Our people's make-up and unique qualities are different from those of the rest of the world. We have Islamic beliefs that constitute a complete and fully integrated system. Free elections are not part of this Islamic system.

(quoted in *The Times* 23 September 1992)

The consequences of Arab governments having so far largely failed to make democracy work have been manyfold. From the perspective of economic development it has meant, on the one hand, that well-intended and appropriate development strategies could not be implemented on the basis of having popular backing, while on the other hand no effective peaceful mechanism has existed to halt even the most obviously misguided of economic measures. In the absence of an open forum in which national development goals could be debated, politicians have instead clung, more perhaps than in other parts of the developing world, to imported ideologies and foreign models of development. As later chapters will show, the landscapes of the Arab world are littered with the failed experiments of both liberal free market economics and socialist ideologies.

It is ironic that one of the forces which may have promoted weak government and hindered the emergence of open political discussion about the desired direction of development has been the easy access to substantial capital of the oil states. In Algeria, Libya, Syria, Iraq, Saudi Arabia and the Gulf oil states capital comes dominantly from state controlled sources (in most cases oil revenues). None of these states therefore depends primarily for its budget on capital raised from the taxation of the population. To some this might seem a desirable recipe for low taxation, but it also means that governments have had neither any political need to be accountable to their peoples nor any incentive to raise incomes as a means of raising taxes. The category of countries mapped as raising most of their incomes from income tax and commodity taxes is more problematic, since many of them (like Jordan, Yemen and Egypt) are economies which are also heavily dependent on foreign aid and the receipt of migrant remittances. Nevertheless it is interesting that the three countries which have advanced furthest towards having politically accountable administrations are also in this category.

To acclaim Tunisia, Jordan and Egypt, and to a lesser extent Morocco, for permitting limited political opposition and therefore to assume that their development strategies are in some way more robust because they have been more openly debated is to attribute a false sense of power to the policy makers of these states. In practice these governments have faced a severe shortage of capital to implement their development strategies (hence the need to raise money through public taxation). The constraint has operated not only to encourage the emergence of limited forms of democratic discussion, but also to seek foreign funding to underpin development plans. This has implied a greater willingness to open their economies to the operation of foreign capital than might otherwise have been the case. The developments which have resulted have not always been desirable or in the long-term interests of their populations. For example the encouragement of international tourism as a means of earning foreign exchange has led to a range of cultural and economic problems for Morocco, Tunisia and Egypt. In spite of these and other problems associated with development strategies dependent on the vicissitudes of international capital flows, it would be fair to suggest that, at least in the short run, these countries have enjoyed the fruits of greater political stability associated with more open systems of government. This in turn has been conducive to investment and economic progress.

One example will suffice to illustrate this point. Egypt was the only Arab country in the 1980s which on average produced more than half of its national demand for consumer goods. This situation was partly a reflection of the investment in heavy industries which took place during Egypt's phase of socialist planning in the 1960s. Sadly in the 1980s Egypt's most successful export earner from manufacturing industry was armaments, sold largely to Iraq to sustain Iraq's war with Iran. In the 1980s public sector industries remained the backbone of Egypt's industrial sector and the government was still having great difficulty in encouraging private capital to invest in Egypt. Lacking domestic private capital, Egypt turned as early as 1974 to seeking foreign capital through the introduction of an open door investment policy. The timing of this policy is significant since it came just after the Yom Kippur War of 1973 and President Sadat's decision to turn to the United States for economic assistance in preference to reliance on Soviet and Arab funds. Sadat's *infitah* (open) economy approach sought to encourage the growth of private industry, first to meet the needs of the domestic market and second to boost industrial employment through the creation of free trade zones, in which industries could be established using Egyptian labour to produce goods for export without paying import–export tariffs.

The signing of the Camp David Peace Accords by the Egyptian and Israeli governments left Egypt ostracized by the rest of the Arab world.

Western investors were less than enthusiastic about using Egypt as a manufacturing export base since it could no longer be used for reaching other Arab countries. At the same time many Arab investors withdrew their funds creating a new economic crisis. The foreign investors who persisted and established export-oriented factories often did so only to exploit cheap female labour and created ventures which were not particularly integrated with the Egyptian economy. In other cases factories were established by foreign investors but without there being much transfer of technical knowledge to the Egyptian labour force, hence minimizing the long-term benefits. By the beginning of the 1980s, Egypt's economic crisis, far from being resolved by its open door policy, had worsened considerably as a result of this range of economic and political circumstances.

President Mubarak, who came to power in 1981 following President Sadat's assassination, continued with the *infitah* policy, but tried to reduce corruption and the bureaucratic inefficiencies which some claimed had hampered its effectiveness (Tripp and Owen, 1989). Initially progress seemed to be slow and investors were very concerned about Egypt's enormous foreign debt which had to be painfully rescheduled several times in the late 1980s. By 1987 non-Arab foreign investors still only accounted for 17 per cent of the private investment projects aimed at producing for the domestic market and 22 per cent of the free trade zone projects. Arab investors, however, had returned and were setting up new export-oriented industries. By the early 1990s the situation had progressed further with the Egyptian government agreeing in 1991 to slowly privatize state industries, despite the political risks associated with the shedding of their over-manned labour forces. Foreign private investors were also showing more interest in the economy because of the government's willingness to step back from protecting the market for goods produced by its state industries. For example, General Motors and Suzuki agreed to expand their car making capacity greatly in Egypt, probably at the expense of lost production by the state car manufacturer. Raising capital to modernize and expand Egypt's industrial sector has therefore required the government to depend upon foreign investment in the economy. This has been at the cost of a considerable loss of control over the direction of Egypt's development path. In place of Egyptian planners and politicians determining the best path for Egypt's industrial future, it appears that transnational corporations and foreign private investors are to be the key actors in the 1990s in determining the country's future.

Conclusion

This chapter has sought to show that political constraints to economic development stem not only from the colonial legacy but also from ongoing problems associated with the exercising of power within the Arab states. Religious and ethnic fragmentation of the populations of the Arab world have combined with serious and ongoing military conflicts to divert scarce resources into the purchase of weaponry and the maintenance of large armed forces. Frequent conflicts have destroyed some of the region's limited infrastructure and the continuing need to support large refugee populations has been a major economic burden for the region. In addition, Arab peoples have been largely frustrated in their attempts to achieve democracy, and where this has been achieved, even in limited forms, elected governments have remained highly constrained in their choices of development strategies because of their dependence on foreign capital to help implement and realize their plans.

Key ideas

1 Religious and ethnic division amongst the populations of Arab countries has served as a peculiarly strong constraint on their political and economic development.
2 The intervention of Western states in many of the conflicts of the Middle East has often aggravated disputes in the region.
3 Arab states, to the detriment of their economic development strategies, have devoted more of their national resources to military expenditure than other developing countries.
4 The Israeli–Palestinian conflict has been fuelled by the immigration and settlement of Israelis on land perceived by the Palestinians to be theirs. Israeli immigration has been mirrored by Palestinian refugee movements and by Palestinian labour emigration.
5 The absence of democratically elected governments in many Arab states has contributed to the weakness of government decision-making on development issues.

Arab oil and the use of oil revenues

Introduction

The analysis of the world oil industry and its implications for the development of the Arab world is an interesting and rewarding topic. No other physical resource of the Arab world is as significant to advanced industrial nations as oil, and no other economic activity reflects as clearly the inescapable interdependence of current events in the Arab world and in the world economy. This is so largely because the Middle East, and the Gulf region in particular, is the location with the largest known oil reserves, believed in 1990 to account for 65.6 per cent of the world total. The geological pre-conditions for oil formation are all present in the geo-synclines of the Gulf (i.e. the presence of oil-forming materials, porous rocks, a sequence of rock strata which permit the separation of oil from water and an impermeable rock cap to trap the oil and prevent it from dispersing). Not only do these conditions occur, but they exist on such a scale as to make oil extraction in the Gulf region more attractive, in economic terms, than almost anywhere else.

Any analysis of the geography of oil involves the examination of four interrelated distributions: oil reserves, oil production, oil consumption and oil sales revenues. Oil reserves relate to the stock of oil believed to be potentially available in the oil-bearing rocks of a country, while only the fraction of those reserves extracted in any one year will enter into the statistics of oil production. For political and economic reasons intimately linked to its development strategy, a country may choose to extract its oil as rapidly as possible or to make it last over a longer period. Oil sales revenues represent the income generated from international sales of that proportion of oil production left over once domestic demands for oil consumption have been met. Some major oil producers such as the Soviet Union and the United States benefit little from international sales because of the high proportion of their oil

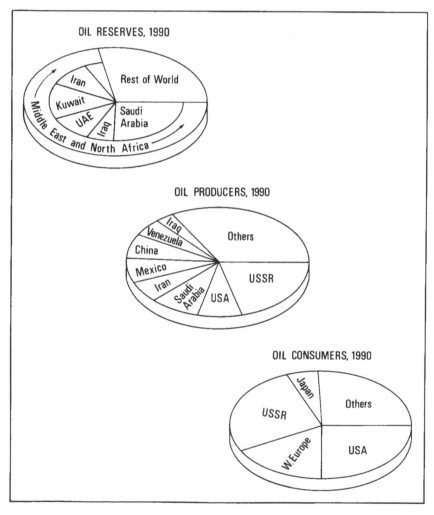

Figure 4.1 World oil reserves, oil production and oil consumption, 1990

production which is consumed domestically. Figure 4.1 illustrates the Middle East countries' position within the world oil industry in terms of these indicators.

Having reviewed the evolution of oil production and trade in the first part of this chapter, attention turns to the way in which oil revenues have been used by the Arab states. Clearly, the income generated has provided the Arab oil states with massive capital reserves to deploy in relation to their disparate, ambitious development plans. Evaluation of the plans leads not simply to a consideration of whether the productive bases of these countries have been able to diversify away from dependence on oil, but more significantly to an examination of what the plans specify as their development objectives. This is an important distinction, since there is a tendency to forget that growth of production in an economy is only a means to an end and not an end in itself. While oil has undoubtedly brought some Arab states great financial assets, the key development issues are how the wealth has been used and which development objectives have been served through the use of the accruing capital.

The organization of the oil industry in the Arab world

The level of concentration which exists in the world oil industry is most apparent in terms of the measure known as oil reserves (Figure 4.1). Although such figures have to be treated with great caution because they are produced by oil companies or oil-producing countries and consequently may be manipulated to serve political or economic ends, it is evident from the data that not only is there a concentration at a world level, but that within the Middle East the five dominant countries are Saudi Arabia, Iraq, Kuwait, Iran and the United Arab Emirates. In 1990 Saudi had a staggering 25.5 per cent of the world's proven reserves, and its Ghawar field had reserves which alone account for more than twice the entire proven reserves of the United States. This high degree of geographical concentration is repeated also at the regional level within the five countries. Oil reserves within these countries are mainly located in the provinces neighbouring the Gulf, as well as offshore. This too has had geo-political implications, since in most cases the oil provinces were not traditionally perceived as forming the core of economic or cultural life of these countries.

Figure 4.1 shows that, in oil production terms, the Middle East is far less dominant than the distribution of oil reserves might suggest. Middle

East oil production in 1990 was much less than in either 1980 or 1970, reflecting the fact that those who control the market for world oil draw on oil sources neither in direct relation to availability nor in direct relation to production costs (both of which would have given the Middle East a higher share). More than compensating for this, the Middle East oil producers consume only a tiny fraction of the oil which they produce, leaving a massive surplus for export, while in many other parts of the world demand for oil greatly outstrips domestic production. For example, Japan used 7.9 per cent of all the world's oil in 1990, all imported, and most of it from the Middle East.

The relationship between global patterns of production and consumption is therefore just as important to the Arab oil producers as is their strong position with regard to oil reserves. The low levels of oil consumption by comparison with the production levels has meant that the Middle East rose to dominate world oil sales as world demand soared from the 1960s onwards. For example, in 1965 25 per cent of world oil exports came from the region while by 1978 no less than 65 per cent of traded crude oil originated in the Middle East. This level of spatial monopoly, both in terms of reserves and in terms of trade, must be virtually unequalled for any other major mineral or fossil fuel resource. The financial resources which have accrued as a result of this particular fact have varied greatly through time and reflect the changing control structure of the oil industry. This topic itself epitomizes many aspects of the ongoing struggle between the advanced industrial and the less developed countries of the world.

As Table 4.1 indicates, the oil revenues received by the Arab states are not directly proportional either to their reserves or to their production capacity. This results from (a) the varying ratio of reserves to production activity as influenced by different state policies on the desirability to increase or limit production in relation to world oil prices, (b) the proportion of oil produced entering international trade as opposed to being used domestically and (c) the inability of some states such as Iraq during parts of the 1980s and early 1990s to make use of its reserves because of war damage to production and transportation facilities, and because of trade embargoes.

Who controls Arab oil?

With such statistics it is perhaps not surprising that the fortunes of the world oil industry have been strongly affected by the policies of a small

Table 4.1 The Arab oil industry: some basic indicators for 1990

	Known reserves (thousand million tonnes, 1990)	Reserves as percentage of world total	Production (million tonnes) 1989	Production (million tonnes) 1990	Export revenues (US$ per capita)
Algeria	1.2	0.9	51.5	54.8	473
Egypt	0.6	0.4	44.5	45.5	n.d.
Iraq	13.4	9.9	138.6	98.2	468
Kuwait	13.0	9.4	81.1	52.7	4,143
Libya	3.0	2.3	55.5	64.7	2,045
Neutral Zone	0.7	0.5	19.3	14.3	n.a.
Oman	0.6	0.4	29.4	32.9	n.d.
Qatar	0.6	0.4	19.3	21.2	4,000
Saudi Arabia	35.0	25.5	257.3	327.1	2,626
United Arab Emirates	12.9	9.7	96.8	109.2	8,792
Others[a]	0.6	0.4	29.1	32.4	n.d.

Sources: British Petroleum, *BP Statistical Review of World Energy 1991*; *Middle East and North Africa Yearbook*, 1991
Notes: [a] Excludes Tunisia which had known reserves in 1990 of 0.2 million tonnes.
n.d., no data; n.a., not available.

number of Middle East oil countries. But there are many other actors in the world oil industry who have intervened in their own interests, relative to trends in oil prices and supply. Figure 4.2 shows some of the more important forces which have been at work. They include the large oil companies, producer-exporting organizations such as OPEC and commodity-purchasing groups such as the International Energy Authority (IEA). These have so far not been discussed, yet they play just as important a part in an understanding of the oil industry in the Arab world as do the policies of individual countries. The significance of these other actors has also changed through time. It is therefore appropriate to consider them relative to five phases which have been identified, by Drysdale and Blake (1985) and others, in the development of the oil industry.

Oil exploration and development: the initial operations of the 'seven sisters'

In the first phase, control of the oil industry, and the benefits accruing from it, lay almost entirely in the hands of a few companies which grew to become the well known oil multi-nationals. Oil was first found in Iran in 1908 by the Scotsman, William Knox d'Arcy. The company which he formed eventually became known as British Petroleum (BP), but its early operation and success arose from very favourable exploration and

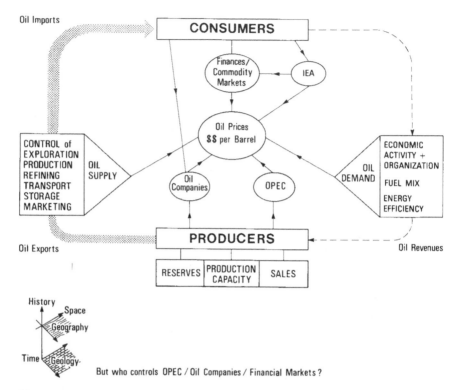

Figure 4.2 Who controls oil? A schematic representation of the world oil industry

production concessions in Iran, which gave BP exclusive control of Iranian oil – a position successfully sustained until 1951.

In the 1920s Shell, Exxon, Mobil and Gulf obtained similar advantageous conditions to explore and produce oil in Iraq. Standard Oil of California struck oil in Bahrain in 1932 and in the next year bought sixty years' exploration and production rights over a large part of Saudi Arabia. The company joined with Texaco, Exxon and Mobil to form ARAMCO (the Arabian American Oil Company), with Saudi oil production commencing in 1937. When oil was discovered in Kuwait in 1938 it was shared on a 50–50 basis by BP and Gulf. Thus the 'seven sisters' were very influential in defining the way in which oil exploration and production began in the Arab world. Their shared power arose from the ability to control

1 the necessary drilling and production technology,
2 the transportation facilities to distribute oil products around the globe,
3 the refining capacity to add value to their oil products and
4 the mechanisms to market oil products to Western consumers.

The fact that there were only a small number of large companies involved in the development of the world oil industry meant that they were able to collude in setting the price of oil and therefore could exercise what has become recognized as one of the classic cases of 'oligopolistic' power. What is often ignored is that this oligopolistic power was evident not just in terms of control of the price of oil but also in terms of an ability, particularly in the Middle East, to control the costs and returns from oil production. The 'seven sisters' under the generous terms of exploration and production concessions were able to develop the oilfields of the Gulf in a much more rational fashion than had previously been the case. Each Gulf oil well was located and developed with the capacity to produce very large quantities of oil, whereas in the early development of the industry in America many small and inefficient wells were owned by companies competing with each other to exploit much more limited underground oil stocks. In the Gulf, therefore, the 'seven sisters', by achieving a collaborative strategy as opposed to operating in a free market competitive situation, were able greatly to reduce their production costs and to raise the returns achieved relative to each well that was drilled. There was not only collusion over the market price of oil but also over the strategy for oil exploration and production.

Growth of oil demand and production: 1945–69

A second phase of development of the Gulf oil industry emerged after the Second World War. World demand for oil and the pattern of production in the Gulf countries changed. Rapid economic growth in the industrialized countries led to a surge in the world's energy needs. While global energy demands doubled between 1950 and 1965, petroleum production tripled. Oil was generally perceived as an efficient, mobile and clean fuel by both industrial and domestic users, and the explosion in private car ownership in the industrialized countries and the diffusion of demand for motor transportation to most developing countries accentuated this increase in demand for oil. This was not just an absolute increase in demand, but a switch relative to the demand for other fuels. It would appear that oil and natural gas accounted for only about one-quarter of world energy production in the mid-1930s, compared with about 70 per cent of the total at the time of peak demand in the early 1970s. This rapid expansion increased awareness in the oil-producing states of the unfair share of the benefits which they were receiving, but it took time for the resulting discontent to be translated into a real power struggle for control of the oil industry.

Prior to the Second World War most oil from the Gulf came from Iran, even though other sources were becoming available just before the outbreak of hostilities. Given this pattern it is not surprising that Iran was first to challenge the oil companies' power. In 1951 a dispute between Iran and BP, following an attempt to nationalize all Iran's oil assets, led to the virtual cessation of oil production for three years. The results of the crisis were profound, including the destabilization of the Iranian economy and government. Not only did the West boycott Iranian oil, but it drove the oil companies to increase production from Kuwait, Saudi Arabia and Iraq, and to extend exploration work in the Arab countries. Within a decade oil was being produced in Libya (1961), Abu Dhabi (1962) and other parts of what became the United Arab Emirates. The conclusion to the first major dispute involving Iran was that in 1954 under a new goverment the industry remained nationalized as the National Iranian Oil Company, but with production and development still being carried out by a consortium of Western companies.

Although in the 1950s concessions granted to oil companies were for shorter time periods, and although profit sharing on a 50–50 basis became the norm, the big oil companies remained immensely powerful. This power base was systematically undermined in the 1960s by two forces. First the oil-producing countries, through the foundation in 1960 of OPEC, succeeded in co-ordinating their efforts to increase their income from oil revenues. Despite income to OPEC countries more than tripling between 1960 and 1970, and despite oil companies increasing their own earnings by more than two-thirds, the remarkable feature of the 1960s was that world oil prices actually dropped from US$1.80 to US$1.30 per barrel. This happened simply because the big companies saw it as in their interest to increase the demand for oil and were more than able to meet that demand through organizing a continuous growth in supply.

The success of the big companies did not go unnoticed. The second force that operated to change the geography of the oil industry in the 1960s and 1970s was the emergence of smaller independent oil companies eager to develop some of the rich newly discovered oil fields. They were able to do so to the delight of these oil-producing countries by under-cutting the terms of the 'seven sisters'. In Libya for example a large number of relatively small independent companies raced to develop the country's oil potential, and succeeded in making it the first Arab state to produce more than 1 million barrels a day. After the closure of the Suez canal in 1967, Libya became a particularly attractive production location, given its proximity to the refineries of southern France and Italy. In some ways the situation in Libya was almost the inverse of that in the Gulf states, with some of the independent producers, such as Occidental, depending almost entirely on Libya for their oil. This position greatly increased the leverage of this oil-producing state relative to the oil companies.

OPEC's golden years: 1970–81

The year 1970 marked a turning point by introducing eleven years of very rapid and sustained increases in oil prices. Continued growth in the demand for oil, combined with a strong dependence by most of the major industrialized nations on OPEC countries to provide oil, enabled OPEC to become highly effective not only in setting world oil prices but in determining levels of oil production. For the first time consumers were presented with the prospect of paying more for their oil, not simply because of increased demand for a scarce commodity, but also because the producers conspired to stabilize or reduce production levels. For once it appeared that some of the less developed countries of the world would be capable of taking the initiative in determining the pattern and pace of their natural resource development.

In 1971, given the tight world oil situation, the 'seven sisters' and six Gulf oil producers met in Teheran and agreed a modest price rise. The instability of the US dollar prompted further price increases within a few months and in 1972 a new agreement was reached which gave every oil-producing country a 25 per cent stake in all oil concessions, scheduled to rise to 51 per cent by 1982. These agreements were soon overtaken by other events, such as the nationalization of BP's oil concessions in Libya in 1971 and of all oil production in Iraq in 1972. This was rapidly followed by Kuwait, Qatar and Iran taking complete control of their oil industries. In Libya, Colonel Gaddafi, who had only seized power in 1969, took advantage of the weak position of the independents. In order to raise prices in a tight world oil situation he insisted that Occidental and other independents cut production levels.

His modest victory inspired the rest of OPEC to force up oil prices. Between 1970 and 1973 they achieved the startling success of doubling the revenue on a barrel of oil. A decisive shift in power within the oil industry had taken place.

The most dramatic action taken by the Arab oil producers came in 1973. Its success would have been impossible had it not been for the pre-existing conditions described above. The consequences of the Arab states using what came to be termed the 'oil weapon' was profound at a global level. The occasion was the Arab–Israeli conflict of October 1973 (the Yom Kippur War). The Saudis called on other Arab oil producers to use their economic power as a lever to counter Western support for Israel. A meeting of Arab oil producers in Kuwait called for production cuts, a sharp increase in prices and a stop to oil exports to supporters of Israel. Of these three measures the key initiative seems to have been the sudden cut in OPEC's production by 25 per cent. This in itself was largely responsible for the quadrupling of oil prices over the following three months. Amongst Arab oil producers, only Iraq did not participate in the production cuts. The Arab oil states enjoyed a rapid increase in the inflow of revenues from oil, and a transformation of their perceived role in the world economy from that of a simple provider of a key energy source to that of a major power broker. The impact of the 1973–4 oil price rises can scarcely be over-emphasized, since they had major implications not only for virtually every aspect of the contemporary economic development of the Arab states but also for the entire world economy.

After the 'oil shock' of 1973–4 there followed a number of years of relative price stability before the next price surge. This was triggered by an oil workers' strike in late 1978, associated with the Iranian revolution. Price rises were sustained through to 1981 by the fear on the world market of the effects of losing 9 per cent of all traded oil. Prices temporarily reached US$40 a barrel. The fact that high prices were sustained in 1980 and 1981 was in some senses surprising, since by mid-1979 Iranian oil production had been restored, but panic buying by traders on the world commodity market turned the expectation of price rises into a self-fulfilling prophecy. Speculative purchases resulted in international inventories of oil growing to more than twice the total volume of oil that had originally 'gone missing' due to the initial Iranian crisis. The outbreak of war between Iran and Iraq in 1980 and the threatened closure of the Straits of Hormuz did nothing to restore confidence. The International Energy Agency (IEA), created after the

1973 crisis to try to pre-empt further oil shocks by building up large stocks of oil, failed initially to recognize that the rise in prices was not the same as a crisis caused by a reduction in supply. It did not therefore release extra oil onto the market to reduce prices. Instead there was for some time an ongoing steady build up around the world of oil stocks, with international dealers failing to recognize that the world oil trade had entered a new era.

One important note should be added to this account, relating to the role of the big oil companies in the 1970s. The actions of OPEC, and of the Arab oil states in particular, had diminished the status of the big companies in the Middle East to that of long-term contractual purchasers, but in terms of their financial interests and those of their owners and share holders the effects were far from detrimental. The profits which they enjoyed from marketing oil at US$15–30 a barrel greatly exceeded those attainable in 1970 when oil had only cost US$1.30. In addition they found themselves wooed by all the major industrial nations and were initially granted remarkably generous concessions to explore for oil in other areas of the world, such as the North Sea.

The erosion of OPEC's power: 1982–5

During the 1970s several important changes took place, which only became evident in terms of world oil prices in the 1980s. First, the significantly higher cost of oil stimulated technological and structural changes in the industrialized countries. Manufacturing companies sought greater energy efficiency, governments introduced incentive schemes to encourage energy conservation, and the public become more energy conscious. The result was that while world energy use had been growing at 4.9 per cent per annum in the post-war period prior to 1973, after the oil price shocks, demand slackened and during the next decade grew at only 2.0 per cent per annum.

Second, the higher price of oil made other fuels more cost effective and also spurred companies to explore for oil in other areas of the world and in higher cost environments. By the 1980s a net reduction in the demand for oil was being registered in the industrialized countries. For example, by 1987 oil accounted for only 44.6 per cent of world energy demand compared with 53.4 per cent in 1979. Furthermore the industrialized countries found that they were becoming more independent of OPEC, since significant oil supplies were now being produced from the North Sea and Mexico. One of the geographical effects of the 1973 price rises was therefore to stimulate a geographically more diffuse pattern of oil production and greatly to reduce the concentration of the international oil trade in the Middle East.

Faced with this reduced demand, OPEC, led by Saudi Arabia, decided to impose a series of production quotas designed to sustain prices rather than to push them up. They perceived the situation as only temporary until oil stocks in the industrialized countries were depleted and until the economic recession affecting some parts of the world economy was over. Recognizing that some non-Arab members of OPEC, such as Nigeria and Indonesia, were less wealthy and less able to sustain production and revenue cuts, Saudi Arabia accepted the role of 'swing producer' making disproportionately large cuts in production to try to stabilize prices. The period was marked by growing tensions within OPEC. Some countries were accused of breaking the quota system while Iran openly refused to accept its quota, given its expensive ongoing war with Iraq.

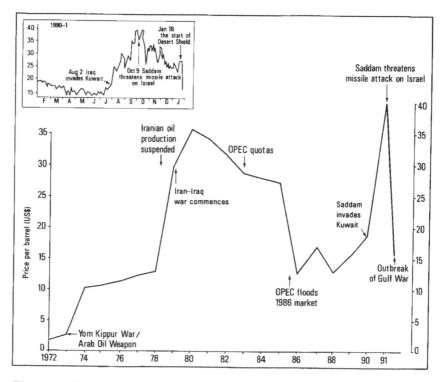

Figure 4.3 The ups and downs of world oil prices, 1972–91

While some seasonal fluctuations in demand for oil certainly were offset by changes in Saudi production levels, the experience of the early 1980s soon showed that the long-term downward pressure on oil prices was not going to disappear (Figure 4.3). Non-OPEC oil producers had effectively eroded the power of OPEC. Also differences of interest within OPEC made it increasingly unpalatable for Saudi Arabia to maintain its role as price-stabilizer. To do so meant Saudi Arabia accepting an ever smaller share of the world oil trade.

Differences in the importance of oil revenues to the development of the Middle East oil states became particularly acute in the early 1980s. Two groups of states began to emerge – the 'spenders' and the 'savers'. The spenders, like Iran and Iraq, had large populations, considerable economic or military commitments, and a strong dependence on further oil revenues to pay off their accumulating foreign debts. Their oil strategy was to maximize short-term income from oil sales by keeping production levels high. The oil savers, like Kuwait, Saudia Arabia and the United Arab Emirates, had relatively small populations to support and found that the rapid surge in oil prices in the 1970s had created for them massive capital surpluses. Oil analysts such as O'Dell have confirmed that the interests of the Arab oil savers was to adjust production quotas and thus oil prices in line with levels of economic growth in the industrial countries. This policy rested on two features: (a) the lucrative prospect for these countries of earning substantial foreign revenues as 'rentier' states from their investments in the industrial nations, and (b) the beneficial production to reserve ratios enjoyed by these countries (on average over 110 years), which pointed to the logic for these countries of sustaining oil's share in the world energy market. For this group of oil producers the logical long-term strategy was therefore to seek a reduction in oil prices during times of world recession in order to maintain oil's share of the world energy market, while gently pushing up prices during times of economic growth in the industrialized countries.

Whether it was a strategy of the 'savers' or a market reaction to a lack of disciplined action within OPEC is hard to tell, but Figure 4.3 certainly shows a continuous decline in oil prices during the first part of the 1980s. In 1985, under the Saudi oil minister's advice, OPEC flooded the world market by pushing up production. World oil prices plummeted, to the dismay both of the oil 'spenders' and of many large oil companies who had invested in expensive oil exploration and production facilities in other parts of the world.

Oil stocks and over-production, 1986–92

Since 1986 there has been a weak world oil market, in terms of both price and demand. The Saudi oil minister's recommendation was subsequently considered a disaster by most OPEC members, but the policy was not necessarily a foolish one from Saudi Arabia's perspective. The short-term effect was to force down the oil price below US$10 a barrel. The subsequent re-introduction of production quotas by OPEC neither resulted in the recovery of the oil price to the levels of 1983 and 1984 nor managed to create stability. Instead price oscillation was extreme, reflecting the vagaries of the international 'spot' market for world oil. The price of oil on the 'spot' market is determined by international speculators whose business is to make profits from buying oil from producers and selling it a few weeks later, after or during the transportation phase.

In these circumstances the greatest benefit accrues to those capable of holding oil stocks close to the final market and releasing oil at the most profitable moment. The fact that many OPEC countries broke their agreed production quotas produced particular uncertainty on the 'spot' market and played into the hands of those controlling oil stocks close to the market. The net result was to transfer power increasingly to the buyers of oil. The term 'buyer' needs to be distinguished from that of 'consumer'. Western consumers have continued to pay high prices for petrol because of the taxes imposed by Western governments and the profits being accrued by those intervening in the world oil trade on behalf of the industrialized countries.

Some of the Arab oil producers have been well aware of the increased importance of getting closer to the consumer in order to increase returns within the oil industry. For this reason in the 1980s Kuwait invested heavily in buying shares in Western oil companies such as BP, while Saudi Arabia greatly increased its floating oil stocks (oil held in its enormous tanker fleet) and enlarged its on-shore storage capacity within Western economies. The oil giant Saudi ARAMCO has bought substantial storage capacity in both Europe and the Caribbean (see Case study F).

Case study F

The growth and development of ARAMCO

The history of ARAMCO (Arabian American Oil Company) encapsulates many important elements in the evolution of the oil industry in the Middle East. In 1923 King Abdul Aziz bin Saud offered a London-based syndicate the exclusive right to explore for oil over more than 30,000 square miles of the Saudi Kingdom. No oil company could be persuaded to provide sufficient funds to proceed with exploration. Ten years later SOCAL (Standard Oil of California) negotiated a new concession to explore in the Eastern Province of Saudi Arabia. The concession was for sixty years and set the royalty price to Saudi Arabia at only US$1.40 for each tonne of crude oil produced. Commercial oil production commenced in 1938, opening a new chapter in Saudi Arabia's economic development.

In 1944 the Californian Arabian Standard Oil Company was renamed ARAMCO. SOCAL subsequently reduced its stake in the company, selling 30 per cent to Texaco, 30 per cent to Exxon and 10 per cent to Mobil Oil. It was not until 1950 that any other changes took place. In this year the Saudi government, in line with other oil producers, renegotiated its concessions and established a 50–50 profit-sharing agreement with ARAMCO. Two years later ARAMCO was persuaded to move its headquarters to Dharan and by the late 1950s two Saudi government ministers were permitted to sit on the ARAMCO board of directors.

During the 1960s the Saudi government sought to establish its own national oil company, Petromin, but given ARAMCO's size and economic resources it was moves in the 1970s towards control of this oil giant which proved much more significant in giving Saudi Arabia power over its own oil industry. Saudi control of ARAMCO advanced in three stages, culminating in a complete takeover of all ARAMCO assets in 1980. Unlike the experience in some other countries the transition from Western to Arab ownership was moderately smooth, taking place over a number of years.

Saudi ownership of ARAMCO was just one stage of a much larger operation to 'Saudi-ize' the oil industry. Expatriate personnel

Case study F (*continued*)

in ARAMCO were replaced, wherever possible, with Saudis. In the 1980s ARAMCO moved to extend its operations through increased investment in the refining and transportation of oil to Western markets. Its subsidiaries have bought a very large tanker fleet, as well as considerable oil storage capacity in Europe and the Caribbean. It has also recently entered into a joint venture giving it control of two of South Korea's larger refineries and retail outlets. These moves to integrate Saudi ARAMCO's operations at all levels in the world oil industry have taken place over a number of years. A new factor since 1990 is that Saudi Arabia has also charged its state oil company with the task of achieving an ambitious expansion programme in terms of its volume of oil production. Production is planned to increase to 10 million barrels per day by 1995. Through further oil exploration in central Saudi Arabia it is hoped that by the year 2000 ARAMCO will be able to pump 12 million barrels per day.

The greatly weakened position of the countries involved only at the production end of the world oil industry has inevitably aggravated the existing difficulties of OPEC, contributing to the collapse in 1990 of discussions between the Arab oil states on production strategies and prices. This was one factor linked with the Iraqi invasion of Kuwait. Iraq, impoverished following its long and fruitless war with Iran, was frustrated with the Arab oil 'savers', especially Kuwait, for their opposition to price-raising strategies. Iraq wanted OPEC to engineer a rapid rise in oil revenues to help pay for its expensive military budget. The oil 'savers', aware of their longer-term interests, saw any substantial rise in oil revenues as counter-productive both to their long-term oil strategy and to their short-term incomes as rentier states.

Oil prices as an indicator of oil power

The previous section of this chapter has provided a chronology of the oil industry in the Arab countries. This has been presented in relation to one important question: who controls Arab oil? It has been shown that spatial control, in the sense of oil reserves being located within the state boundaries of a country, does not ensure that a producer country benefits to the maximum from oil sales. Of much greater significance are the intervening actors in the world oil trade. Through time the relative power of these different actors has changed significantly, in relation to each other and to the varying world demand for oil. Perhaps the most reliable indicator of who at any one time controls oil is the oil price as plotted in Figure 4.3. Major gains were achieved for the Arab oil states as a result of OPEC's actions in the early 1970s. The increase in aggregate oil income of OPEC members from US$2.3 billion in 1960 to US$90.5 billion by 1974 has justifiably been described as one of the most spectacular short-term wealth transfers in the world's economic history. The price gains of 1979–80 reflected continued pressure of demand relative to supply, but these gains resulted as much from the operation of international commodity markets as from the actions of OPEC. Since the early 1980s it has not been OPEC so much as the interests of Saudi Arabia and the other Arab 'saver' states that has been critical on the supply side of the oil trade.

One of the most remarkable features of Figure 4.3 (see inset) is that, on the day in January 1991 when the 'allied' military offensive against Iraq's invasion of Kuwait was launched, world oil markets experienced the biggest single day price drop in history. Before the war it had been predicted by some, who still based their thinking on the circumstances of 1973 and 1979, that an outbreak of hostilities would triple oil prices because of the threatened disruption of supplies. That the opposite happened was no accident and reflects the changing balance of power which has occurred in the world oil trade.

Before the invasion oil prices oscillated around the US$10 level. Iraq's invasion of Kuwait in early August 1990 led to an immediate speculative rise in prices of the same kind as had been experienced in 1979, so that by September 1992 prices passed through the US$40 a barrel level. The world boycott of Iraqi oil had of course trimmed 4.5 per cent of world oil production from international trade, while the loss of Kuwait's 2.6 per cent of world production further reduced oil supplies. Added to this the prospect of a war affecting the Saudi oil fields introduced all the elements for a real oil crisis and helped to explain why speculative buying pushed up prices so fast. However, long before the armed hostilities actually began prices were falling again, indicating who were the real power brokers in the contemporary oil market. These proved to be neither Saddam Hussein, who in a geographical sense was in spatial control of considerable reserves, nor those OPEC states seeking to raise prices as a result of the temporary crisis in oil supply. The Arab oil states allied against Iraq chose to increase their production levels markedly, with the result that by late 1990 the oil-producing states were pumping 23 million barrels a day for a world demand of about 21 million. In addition, as the date for hostilities approached, the IEA chose to release from its massive stocks about 2.5 million barrels a day, thus creating over-supply at the very time when speculative market forces might otherwise have produced exactly the opposite effect.

This joint action by Saudi, as the world's largest oil producer, and of the IEA, as a major regulator of oil supplies operating in the interests of oil consumers in industrialized countries, was of course of limited duration, but it indicated that they are key actors in affecting oil prices. Although prices for oil will continue to fluctuate in relation to the strategies of the key actors, it would appear that in the near future those with the greatest leverage over oil prices have a shared interest in maintaining economic stability in the industrialized countries. The history of Arab oil teaches above all else that the future is unpredictable. A change of political regime in Saudi Arabia or a switch in the quality or quantity of world energy demands or a split in the interests of the industrialized nations behind the IEA could unexpectedly introduce yet another transformation in the geography of Arab oil.

Infrastructure development in the oil states

The way in which the oil industry evolved permitted governments or rulers of the Arab oil states to become both the recipients and the disbursers of oil revenues. The non-oil states indirectly benefited from oil through their involvement as suppliers of migrant labour, and as recipients of aid and loans from the oil states. The latter theme is taken up in Chapter 5.

The rapid rise in oil revenues in the oil states produced an era of unprecedented infrastructural development. Investment in improved physical infrastructure was welcome in all the Arab oil states since it improved the quality of life of their populations, but the extent to which it served to add to these countries' long-term development prospects was variable. Building new roads, airports and harbours in virtually all the Gulf states was initially of economic value, since it reduced bottlenecks in importing and distributing goods within their space-economies, and hence assisted in reducing inflationary pressures. Unnecessary investment in duplicate facilities, and investment in communication networks for which there was no long-term demand, occurred in the late 1970s, and at best reflected a confusion in development planning between investment in physical infrastructure as a means to achieving development objectives and the objectives themselves.

Port development around the Arabian peninsula in the 1970s and 1980s is one example of infrastucture investment being viewed as an *end* product of the development process rather than as a *means* to an end. Oil wealth led to two types of imports being demanded on a new scale in the oil states. First, the launching of ambitious new building programmes boosted demand for construction goods and the necessary equipment to progress with the planned development projects. Second, the increased wealth of the populations of these states fed an apparently insatiable demand for imported consumer goods. The result was a vast increase in pressure on the ports of the oil states. This caused a sudden increase in the need for more berths and port facilities in all the Gulf states and in Saudi Arabia's Red Sea ports. The inadequacy of existing facilities was evident in the long delays experienced by shipping. For example in the mid-1970s a delay of 150 days was not unusual for ships waiting to dock at Dammam. Port bottlenecks slowed delivery of basic materials such as concrete needed for infrastructure development.

Between 1978 and 1985 the number of berths in all the Gulf states was expected to increase from 114 to 397. This rapid expansion effectively overcame shipping delays by the 1980s and instead a problem of over-capacity had emerged. Why had this reversal occurred? First, there was a drop in demand for construction goods by the mid-1980s due partly to the completion of the early construction phase of development and partly to the newly established capability of some of the oil states to produce their own concrete and other construction materials. Second, by the mid-1980s the drop in oil prices had reduced demand for consumer imports. Competition to attract traffic became a serious matter in the newly expanded port system of the Gulf. Over-capacity was most obvious in the ports of the United Arab Emirates with their small hinterlands. The 1980s saw the smaller ports lose out to the larger ones. Ports such as Dubai, where some 42 per cent of imports at one point had been cement, were particularly severely affected by the downturn in demand. Other ports attempted to specialize in service functions. Bahrain, for example, moved to become a centre for ship repair and servicing in the Gulf. The Iran–Iraq war also affected shipping patterns, with trade being diverted from the war zone and with some ports gaining a transit-trade function. Aqaba in Jordan became an important transit port for Iraq, while Khor Fakkan and Fujairah attracted shipping companies not wishing to risk entering the Gulf.

Transport infrastructure developments brought considerable social and cultural changes. New roads and modern public transport services have linked together peoples and places which until a few decades ago had remarkably limited, and certainly strictly constrained, contact patterns. For example, the completion in 1986 of a new 24 kilometre causeway between Saudi Arabia and the island of Bahrain not only altered the retailing structures and the property market of Bahrain, but also offered to Saudi citizens of the eastern province (used to the strictly enforced Islamic norms of Saudi society) easy access to a small and socially much more liberal state (Unwin, 1988).

Oil revenues were used by the Arab oil states not only to create a better transport infrastructure but to expand other aspects of economic, social and political life. This meant the building of large numbers of schools and colleges, hospitals, clinics and other social utilities. Much more publicity was of course given, in countries such as Saudi Arabia, to investment in health and education programmes than to the 20 per cent of GDP devoted to the development of military infrastructure. Ironically, both public services and military expansion implied a growing dependence on external technical assistance. Not only were oil revenues disbursed to establish this new infrastructure, but future oil revenues were also implicitly committed to these developments to maintain such facilities as modern hospitals and sophisticated military systems using Western technlogy. Infrastructure development cannot therefore in every case be interpreted as laying a sound base for future economic and social progress. In some cases it implied a commitment to a future development course which was dependent on external technological assistance and which would require the diversion of future oil wealth to pay for this, rather than for other programmes.

In the 1970s the vast scale of oil revenues in countries such as Saudi Arabia, Kuwait, the United Arab Emirates and Libya permitted ambitious projects to advance and at the same time permitted their governments to accrue substantial budget surpluses which could be used for other purposes. In Iraq and Algeria as well as in other populous oil states outside the Arab world ambitious development projects were launched in the 1970s on the assumption that oil price increases were permanent and would be sustained. This led to these states borrowing capital on the strength of their anticipated future oil earnings to launch their development programmes. The appropriateness of such development plans has to be judged, not only in terms of the wisdom of the way in which current oil revenues were being disbursed relative to the needs of their populations, but also in terms of the longer-term consequences of major foreign debt repayment.

The Libyan path to development in the 1970s and 1980s

Libya provides an interesting example of one country which found that, while reducing oil production levels in the 1970s, it was able to expand its oil revenues and to launch a remarkable range of new, so-called, 'development' projects. The explosion of oil revenues after use of the 'oil weapon' meant that by 1974 the oil industry accounted for 61 per cent of the country's GDP. A further 25 per cent was related to tertiary activities, mainly supported by the disbursement of funds by the government based on its oil wealth. These funds were allocated to education, health and other service sectors. Some 10 per cent of GDP financed the construction industry creating a better public infrastructure and improved housing. In 1974, only 4 per cent of the Libyan economy could be said to have been generated by non-oil-related activities such as agriculture, mining or manufacturing.

In both its 1973–5 and its 1976–80 development plans Libya chose to disburse 58 per cent of expenditure to public services or to construction programmes associated with future public utilities. Agricultural projects were launched such as wheat cultivation schemes at Kufra and elsewhere, based on the mining of deep aquifers to feed oasis agriculture. Since only 1.1 per cent of Libya's land area permitted arable agriculture, the potential for sustainable agricultural development was always highly constrained by the country's extreme water deficit. Even in the Jifara, the coastal plain of Tripolitania, and in the northeast in the Benghazi region, investment in new irrigation schemes was often ill-founded, involving increased water extraction rates from the local aquifers to six times the rate of recharge. This resulted by the late 1970s in substantial sea water incursions and hence a reduction in the quality of the water available. It was in this context that Libyan planners began to contemplate yet further extravagant investments in water 'mining' and transportation from the south of the country.

Investment in industry in Libya in the 1970s was also problematic. The country had inherited virtually no industrial base from its period as an Italian colony. Apart from oil it had few mineral resources. Its very small population did not provide a large enough domestic market to encourage the development of manufacturing. In addition the very limited stocks of water, which proved a constraint to agricultural expansion, also restricted the potential for many types of industrial venture, as did the scarcity of labour. From 1969 onwards Colonel Gaddafi sought a socialist path to Libyan development. By nationalizing foreign banks, confiscating all remaining Italian properties and pressurizing oil companies, he limited the courses for industrialization, both in terms of his ideological objectives and in terms of the international alienation triggered by his policies.

The 1976–80 plan allocated 15 per cent of its budget to industrial projects. Some of this was inevitably destined for petroleum-related activities at Masra al Brega, Misurata and Ras Lunaf. This investment, while having some economic logic in seeking to add to the value of Libya's oil exports, did little to reduce the economy's dependence on oil. Other funds were disbursed on prestige projects symbolizing Gaddafi's desire to achieve Libyan self-sufficiency. An iron and steel complex was built with Western technological assistance on the coast at Misurata. This was ill-conceived, requiring imports of iron ore and of foreign skilled staff, yet needing foreign markets to achieve efficient production levels. A similarly ill-founded project was the scheme for an aluminium smelter near Tripoli. Prestige projects such as these boosted the country's already soaring import bill for equipment, manufactured goods and foreign services. The geographical effects of the pattern of the Libyan government's planned expenditure was to further concentrate economic activities in the two coastal areas around Tripoli and Benghazi. It was to these cities and neighbouring coastal settlements that most population migration occurred in response to the spatial concentration of investment in these zones.

When oil revenues dropped in the early and mid-1980s, Libya, like the other oil states, had rapidly to adjust its expenditure plans. Many of the over-ambitious projects begun in the 1970s (such as the building of an enormous artifical river to transport water from four large reservoirs in the south to the populated coastal strip) had to be either cut or postponed. By 1988 Libya was experiencing a deficit in its financial balance. By 1990 it was evident that the huge investments in agricultural and industrial projects of the 1970s had either failed or were yielding relatively poor results. In addition, the United States, aware of Libya's dependence on Western technology and frustrated by Libya's support for anti-American groups, imposed a commercial boycott, making it difficult for the country to sustain its remaining development objectives. The costly $3 billion water pipeline from the south to the

Gulf of Sirte had by the mid-1980s become linked to the objective of simply sustaining water self-sufficiency in Libya's coastal urban agglomerations. Experts have for some time pointed out that for Libya the goal of food self-sufficiency is no more realistic an objective, on the basis of its very limited water and manpower resources, than is the goal of industrial self-sufficiency on technological, resource and market grounds (Allan, 1985).

Agricultural and industrial development in the Gulf states

The development programmes adopted by the Gulf oil economies differed from those of Libya in several respects. Their political ideologies were not inherently opposed to individual enterprise and to capitalist modes of production: there was a greater willingness to permit direct foreign involvement in development projects and there was also some encouragement by the state for private sector initiatives. Investment of Arab oil revenues in the industrialized West was one mechanism for diversifying future income sources away from the oil sector.

Saudi Arabia, Kuwait and the United Arab Emirates invested in oil-related industries such as fertilizers and petro-chemicals. Joint (Arab–foreign) industrial projects were launched in a range of sectors from food processing, through building materials, to downstream oil-related products such as plastics.

Only toward the end of the 1970s did private investors begin to take real interest in the substantial subsidies made available in Saudi Arabia for agricultural development. The government was willing to give land to any farmer with an appropriate scheme to help the country move towards food self-sufficiency. In addition various gifts, subsidies and price support mechanisms were made available to farmers to invest in buildings and equipment, to drill wells, to buy seeds or fertilizer and to sell crops. By 1982 Saudi Arabia was producing 240,000 tonnes of wheat per annum, about half of it from farms in the Qassim province north of Riyadh. The whole drive to expand agricultural production was strongly encouraged by Western companies and agricultural consultants eager to provide equipment and technical expertise.

In Qassim wheat cultivation was based on centre-pivot irrigation systems. Very deep wells going down more than 300 metres tapped aquifers charged with water thought to be 35,000 years old. While this form of irrigated wheat cultivation stamped green circles on the face of the Saudi desert, the inevitable question, as in Libya, was how long these aquifers could be tapped, since extraction rates greatly exceeded recharge levels. During the 1980s the water table dropped by more than 100 metres in the Qassim area and deeper drilling was producing hotter, saltier water.

Unlike Libya, Saudi Arabia could tentatively claim some success in achieving agricultural self-sufficiency in some crops by the 1990s. Soaring production levels of wheat turned Saudi Arabia into a net wheat exporter during the 1980s. In 1990 the government paid US$2.1 billion in wheat subsidies for the record harvest of more than 4 million tonnes. This was more than three times domestic requirements and sale of the surplus made the Kingdom the world's sixth largest exporter of wheat. By 1990 Saudi Arabia was also self-sufficient in eggs and certain fruit and vegetables and close to self-sufficiency in dairy produce.

The cost of this type of agricultural development in environmental and economic terms is very great. Studies suggest that at current rates of water extraction fossil water supplies in the main wheat-producing areas such as Qassim will be exhausted within ten to twenty-five years. Some of the early farms in Qassim had already been abandoned by 1990 and problems of soil erosion were being reported. The slump in oil revenues in the mid-1980s also forced the Saudi government to reconsider its subsidies to agriculture, given the much greater cost of producing most agricultural produce than of importing it. The problem for the government became that of discouraging farmers from producing surplus crops which Saudi Arabia did not need and therefore did not want to subsidize. Critics of the government's agricultural policy suggest that exporting wheat is effectively no more sensible than Saudi Arabia exporting water. The policy of seeking self-sufficiency in dairy produce has come under similar criticism. It has been calculated that for Saudi Arabian farmers to produce one gallon of milk requires on average 1,500 gallons of water to produce fodder crops for the cattle, to water the cows and to sluice the animals to keep them cool. Why therefore has self-sufficiency in the agricultural sector been so widely adopted as a development goal in Saudi Arabia and in the Arab world as a whole?

Unlike agriculture, the potential for industrialization in Saudi Arabia and the Gulf states has one great strength: the cheap availability of oil as an energy source. The circumstances for rapid industrialization in Saudi Arabia were also more propitious than in Libya. From the outset Saudi Arabia sought to establish a very specific type of industrial base. The industries which were encouraged were to be highly automated and to have low water requirements to offset the two main weak points of the economy. The thrust of the industrialization programme was located in just two cities: Yanbu on the Red Sea, and Jubail on the Gulf. These locations (chosen in 1976) sought to keep new developments away from existing population centres, so that they would not compete for access to local water resources. The sites became the base for chemical industries and oil refineries at the termini of the new 600 mile east–west oil pipeline. Foreign technology and markets for chemical and refinery products were assured by developing joint industrial ventures between SABIC (Saudi Arabia's Basic Industries Corporation which was created by the state to add value to Saudi's hydro-carbons) and foreign companies. By the end of the 1980s SABIC's fifteen industrial plants were earning handsome profits from the export of petro-chemicals mainly to the Far East.

By the time of the Fourth Saudi National Plan (1985–90) the government felt that all the physical infrastructure for substantial development had been established and that state investment in basic industries should now be accompanied by more private investment in secondary industries. Estates for private industrial ventures were built at Yanbu and Jubail to accommodate factories to manufacture plastics, fertilizers, fibreglass and other goods. In search of free enterprise involvement, the Saudi Industrial Development Fund offered attractive loans, tax concessions and cheap industrial sites. By far the three biggest concentrations of light industry were in Riyadh, Dammam and Jeddah. This was not surprising since these cities controlled the main domestic markets for manufactured goods.

Figure 4.4 shows the industrial geography of Saudi Arabia in the mid-1980s in terms of the distribution of industrial licences. Of course not all industrial licences resulted in successful manufacturing projects. It should also be remembered that industrial projects vary greatly in size and value. Nevertheless Figure 4.4 provides an indication of the pattern of industrial development which has emerged in the Kingdom. Many activities related to the construction sector were badly hit when oil prices slumped and infrastructural and building projects were cut back.

In the mid-1980s two events associated with the crash of oil prices actually helped manufacturing industry to expand in a more sustained fashion. The first was that the Fourth Saudi Plan threatened greatly to reduce the size of the expatriate workforce. In practice no major exodus of migrant workers occurred prior to 1990. Instead wage levels were forced down, making it cheaper for Saudi Arabia to employ both skilled and unskilled immigrant workers. For the emerging industrial sector this

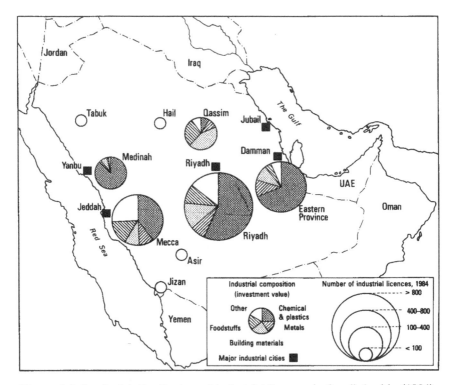

Figure 4.4 Provincial distribution of industrial licences in Saudi Arabia (1984) and the value of planned industrial investment

made labour costs more competitive on an international scale. Second, the Saudi government, faced with cuts in oil revenues but still wishing to purchase foreign technology and military hardware in view of the perceived threat that the Iran–Iraq war might spread, engaged in some 'offset deals' with foreign governments. They agreed to buy expensive (military) systems from the West on the understanding that the expenditure costs of these deals would be *offset* by Western companies' investments and purchases from Saudi Arabia. Thus, for example, British Aerospace entered a number of joint ventures in Saudi Arabia as part of the British offset agreement (the al-Yamamah pact) signed after Saudi Arabia's major purchase of British Tornado aircraft.

Examples of the type of industrial developments which resulted included British Aerospace's involvement in the establishment and running, in partnership with Saudi entrepreneurs, of an aluminium smelter in Yanbu. Another joint venture with Metal Box resulted in Saudi Arabia becoming a major manufacturer of cans from a factory in Jeddah. The latter project was highly appropriate, given the large market for soft drinks in Saudi Arabia and the neighbouring Gulf states. In total these states have a small but wealthy combined market of about 18 million people. There is therefore some limited evidence of success-ful industrial development in Saudi Arabia arising out of the disburs-ment and re-investment of oil revenues, but the question remains to be answered as to whether this is the most appropriate form of development. First, there is a reluctance by many Saudi men to work in industry and, in the absence of female entry to this part of the labour market, a continuing need for an expatriate workforce. Second, this development path implies an ongoing dependence on Western technology and managerial skills.

In summary, industrialization has been pursued as a key aspect of the development strategies of most of the oil-rich states. The logic in Saudi Arabia and the other Gulf economies has been to encourage economic activities which are in harmony with the region's natural production advantages – low capital and energy costs. The objective in these countries appears to have been not so much a lessening of dependence on foreign technology, manpower or trade, but a diversification of future revenue sources away from oil sales.

Kuwait: the emergence of a rentier state

Industrialization has not been the only strategy to offer a reduced dependence on oil revenues. Limited investment opportunities combined with rapidly accruing oil revenues led countries such as Kuwait to divert investment to Western industrialized countries. As early as 1961 Kuwait established the first of its three major investment companies (KIC, the Kuwait Investment Company). In 1964 the Kuwait Investment Office (KIO) was opened in London to handle dealings on the London market. In the 1970s as oil revenues soared, tangible low-risk and long-term investments were being made in many leading Western companies. In 1976 the Reserve Fund for Future Generations (RFFG) was established to manage foreign investments arising from allocation to the fund of a 10 per cent share of Kuwait's annual oil income. By 1982 Kuwait was reported to hold stock in no less than 480 of the top 500 companies in the United States as well as a portfolio of major holdings in British institutions such as shares in the Bank of Scotland, Guardian Royal

Exchange and General Accident. Real estate was purchased around the globe including hotel chains, highland estates and islands off the American coast line. Kuwait became a major share holder in established car empires such as Volkswagen and Daimler-Benz and less surprisingly took a 20 per cent stake in BP as well as in the European distribution network for Gulf Oil.

As a result of this foreign investment Kuwait, especially its rulers, began to receive significant income from overseas investments. By 1982–3 the RFFG, whose income was not meant to be touched until the year 2001, was already generating $9 billion worth of income compared with Kuwait's oil earnings of $12 billion. It only took oil revenues to drop in the mid-1980s for investment earnings to overtake oil as the main source of income and to establish Kuwait as the first 'rentier' economy in the world. The creation of a Kuwaiti stock exchange was not, in the context of the country's investment strategy, particularly surprising. Its crash in 1983 temporarily shook confidence in the system, but was far from the most fundamental problem of Kuwait's rentier policy. More problematic was the lack in Kuwait of other strong development policies which might have distributed oil wealth to the population more usefully (e.g. through industrial jobs). Instead the government kept living standards high for its citizens during the recessionary period of the mid-1980s by a system of subsidies and state social benefits allocated on a restrictive basis and not shared to any significant extent with Kuwait's large immigrant labour force.

Less publicized but still significant has been the foreign investment policy of the United Arab Emirates. This has followed a similar path to that of Kuwait. Nor should this mode of capital expenditure be thought of as only affecting the smaller oil states and consequently unimportant. Saudi Arabia has also invested heavily abroad, although it has been less concerned to hold long-term assets. The scale of the flow of capital out of the Arab oil states into bank deposits in the Western world was estimated by the Bank of England to be of the order of one-third of the oil exporters' surpluses in the 1980s. Obviously these countries were not relying on the success of domestic investment as the sole, or even the main, source of future revenues. The declining value of the dollar in the mid-1980s and the experience of industrial recession in the West only served to encourage a geographical diversification in investment patterns so that the portfolio of Arab investments became more truly global in nature.

In development terms the richer Arab oil states of the Gulf, with the exception of Iraq, by the early 1980s had tied their own future development courses to participation in the world economy. This had been achieved in each country in slightly different ways, as has been illustrated in this chapter. To a lesser or greater extent all the major oil states (except Iraq which was distinctive in a number of ways) looked externally to the world economy for access to technology, for markets for their oil and non-oil products and as a location of investment for their capital surpluses. For the wealthy Arab countries there was a growing interdependence of interest with the world's more industrialized economies.

Development paths and development objectives

The latter part of this chapter has reviewed the diverse development paths followed by some of the Arab oil states. The terms 'diversification', 'self-sufficiency' and reduced 'dependence on oil' have recurred throughout the discussion, making it desirable to conclude this chapter by seeking to relate these terms to what might be considered to be the objectives of economic development in these states. All too often concepts like 'self-sufficiency' are confused in policy makers minds with what might be considered to be true development objectives. Bowen-Jones (1984), from a survey of the national development plans of the Gulf region, has suggested that five true development objectives operated in the 1970s and 1980s:

1 maintaining current high levels of material consumption in a post-oil era;
2 maintaining respected cultural and social values and stability;
3 food security;
4 greater satisfaction with the quality of life and the environment;
5 a stronger command over regional and national destinies.

These provide interesting yardsticks against which to measure the development paths which have been followed. In the light of these goals it soon becomes clear that policies aimed at reducing dependence on oil select this path to achieve rather more important underlying development goals, such as the first one listed above.

It has been shown that other policies such as self-sufficiency in food cannot sensibly be attained in all Arab countries; even where food security has been achieved in the short run, it has often been at the expense of diverting scarce water resources, hence involving the risk of a lowering of the quality of life in the longer term. Most would expect prudent expenditure on physical and social infrastructure to raise the quality of life of the citizens of the oil states but, as with the Saudi–Bahrain causeway, this has not been achieved without threatening pre-existing cultural and social values. The building and maintenance of infrastructure and industrial projects has also involved the introduction of a large expatriate labour force. This in turn produced the geographical juxtapositioning of people from different cultures with all the risks that this has had for a dilution of Islamic values. It is interesting in the aftermath of the Iraqi invasion of Kuwait that one of Kuwait's most forceful new policy objectives has been to re-establish Kuwaitis as the majority population group within the state. This goal is likely to be achieved by a reduction of all categories of foreign workers, but especially of other Arabs.

In reviewing the development achievements of the oil states over the last two decades from the perspective of the 1990s, it would seem that a confusion has arisen between the means of achieving particular ends and the long-term goals of development. This confusion has cost several countries dearly. Use of oil revenues in Libya over the twenty years would seem to have advanced it little in terms of achieving the fifth goal listed above. By contrast, Kuwait's foreign investment policy yielded considerable financial security when its oil assets were seized, but its lack of a strong internal development policy to disburse its wealth more equitably threatened its long-term stability and meant that significant proportions of the non-Kuwaiti workforce sided with Saddam Hussein when he invaded the country.

Conclusion

The wealthy Arab oil states provide an interesting focus for the study of the development process. Unlike most developing countries, lack of capital has not been a major constraint on their choice of objectives or development path. The abundant availability of capital has not proved a sufficient condition for them to enjoy a smooth development course. Capital surpluses are only of value when appropriate development paths (in economic, ecological and social terms) are chosen which are in harmony with clearly defined development objectives.

Key ideas

1 The development of oil production in the Arab world is integrally linked to the changing organization of the world oil industry. The governments of oil-producing states, multi-national oil companies, OPEC, the IEA and speculators on the world commodity markets have all been influential in varying ways and at different times in shaping patterns of oil production.

2 Oil prices act as a useful indicator of oil power. In the 1980s and 1990s the Arab oil states have found it more difficult than in the 1970s to influence the world price of oil.

3 Arab oil-producing countries may be classified as 'spenders' and 'savers', reflecting fundamental differences in their long-term strategies on oil pricing and production.

4 Oil revenues earned by the Arab oil states have been a major catalyst to economic and social change. Development strategies using oil revenues have ranged from Libya's search for industrial and agricultural self-sufficiency to Kuwait's economic diversification through becoming a 'rentier state'.

5 Analysis of development plans suggests that true development objectives, such as achieving a stronger command over regional and national destinies, have often been confused with the means of achieving these objectives.

5
Labour migration

Introduction

Although many Arab countries have no oil, there are few which have not experienced some major economic consequences of being located close to the centre of the world oil industry. Some have gained income as countries through which oil is transported (as in the case of Egypt with its revenues derived from the passage of oil tankers through the Suez canal). Others received large in-flows of remittances sent by migrants working in the oil states. In the 1980s in Yemen and Jordan, migrant remittances were worth many times the value of exports, and remittances were valued at the equivalent of between a quarter and a third of their GDPs. Income sent home by migrants working in the oil states has played an important role in shaping both the labour market and urban development patterns in Arab labour-sending countries. A dependence on the world oil industry has not therefore been an issue only for the Arab oil states, but has been a concern shared by the other Arab states.

Unlike most developing countries, the Arab oil states found their development plans strongly constrained by lack of labour and of specific skills rather than by lack of capital. In the 1970s most Arab oil producers sought to overcome this problem by introducing large quantities of immigrant labour. By 1975 almost half (48.7 per cent) of the workforce of the Arab oil states (excluding Iraq) consisted of foreign workers. The number of labour migrants and their families grew in the decade which followed from an estimated 2.8 million in 1975 to 7.2 million in 1985.

So how has labour migration evolved in the Arab world over the last two decades, and in particular how has it affected the economies of the labour-sending countries? The development implications have been substantial since international migration involves the transfer between states of labour resources. While one state paid for the migrant's education and training another state gleaned the benefits of the productive value of the migrant's skills. The gains and losses to migrant-receiving and migrant-sending countries are much more complex, however, than simply the lost returns on investment in human training (sometimes referred to as human capital). Associated with most labour migration flows there are related flows of migrant families, the return movement of migrant savings (remittances) to the country of origin, inter-state technology transfers and the international movement via the migrant community of information about political, economic and cultural systems.

Migration and development in labour-sending states

In the context of the Arab world, migration has triggered a wide range of development problems in a particularly acute fashion, simply because the labour-sending countries of Yemen (formerly the People's Democratic Republic of Yemen and the Yemen Arab Republic), Jordan and Egypt are arguably the most 'migrant dependent' countries in the world (Table 5.1). This claim of extreme migration dependence is not examined in great detail here, but is based on the author's research using indicators such as the proportion of a state's adult labour force who work abroad and the ratio of migrant remittances to the value of domestic economic production (GDP). For example in the Yemen Arab Republic between 1980 and 1985 migrant remittances were worth over four times the value of the country's exports. Table 5.1 shows the six countries in the world with the highest ratio of net remittances to exports. As can be seen for the years 1980–5, four of the six top countries were Arab states, and the other two, Pakistan and Bangladesh, were directly involved in sending workers to the oil states.

What the indicators shown in Table 5.1 imply is that the Arab world in the 1980s was the chief locus of the world's most significant labour migration system. The table also shows the effect of declining migrant wages on remittance income in the latter part of the 1980s in Jordan and Egypt. For some countries (the oil states) economic development plans were contingent on the continuation of massive stocks of

Table 5.1 The world's most migrant dependent states

	Remittances relative to value of exports[a]		Remittances relative to value of GDP		
	1980–5	*1986–9*	*1984*	*1987*	*1989*
Yemen Arab Republic[b]	4.8	6.4	0.23	0.24	n.d.
People's Democratic Republic of Yemen[c]	3.9	n.d.	n.d.	n.d.	n.d.
Jordan	3.4	2.1	0.21	0.13	0.10
Bangladesh	1.5	0.9	0.04	0.06	0.05
Pakistan	1.0	1.0	0.11	0.11	0.08
Egypt	0.7	0.3	0.21	0.08	0.04

Source: derived from International Monetary Fund (1991) *International Financial Statistics Yearbook 1990*
Notes: [a] Exports are defined as goods and services.
　　[b] Export data only available for 1986–7. Thus the main downturn in remittance income is not apparent for the Yemen Arab Republic in this data set.
　　[c] Data only available for export of goods.
　　n.d., no data.

immigrant workers. For other states, income from their citizens working abroad as migrants was a dominant part of their economic structure. If dependence on migrant remittances produces distinctive forms of economic development, then it would be countries such as Yemen, Jordan and Egypt that one would expect to reveal the characteristics of such a situation. The latter part of the chapter therefore turns to the issue of what extreme migration dependence has meant for the development of these states.

The shifting demand for immigrant labour

Table 5.2 shows the scale of labour immigration to the main oil-producing countries of the Arabian peninsula. Labour availability acted as a constraint on the implementation of development plans in these countries for several reasons.

First, with the exception of Iraq, the Arab oil-rich states had very small populations which could never have met the labour demands of the development programmes adopted by their rulers as they sought to transform their newly gained oil wealth into tangible benefits for themselves and their populations. For example, in 1975 Saudi Arabia was estimated to have a national workforce of just over 1 million people. Like other Arab countries the activity rate of the adult population was low because of the very low participation rate of women

Table 5.2 Estimated stock of immigrants in certain Arab oil states[a]

Country	1975	1985	Percentage active to Arab immigrant population 1985
Bahrain	56,000	121,800	45
Kuwait	502,000	1,016,000	39
Oman	132,250	391,000	43
Qatar	97,000	129,200	23
Saudi Arabia	1,565,000	4,504,700	71
United Arab Emirates	456,000	1,038,800	36
Total	2,808,705	7,201,600	

Note: [a] Estimates of migration stocks vary widely. The figures in the table are based on Birks and Sinclair (1989, 20–1).

in the workforce. Second, prior to the arrival of oil wealth, the labour forces of these countries had not received the training and skills necessary to launch and maintain technologically advanced agricultural and industrial systems. Investment in educational infrastructure to remedy this shortfall rapidly became an objective of many of the oil-rich states. However, training and skill acquisition involves long-term investment and so the needs of these development programmes could not hope to be met without labour immigration. Third, the workforces of these countries have often proved resistant to entering certain occupations. One of the unfortunate side-effects of the improved welfare services for the indigenous populations was a reduced incentive to take up jobs perceived as 'undesirable' or of low status. In addition the policy in many Arab oil states of guaranteeing all university graduates a position in a government or state-run organization has been to reduce the proportion of nationals in the private sector and to increase the scale of immigration.

The shortages of supply of national labour, both in terms of quantity and quality, led to labour immigration. The pattern and evolution of the ensuing labour flows into the oil-rich states has sometimes been represented as the classic case of migration responding to a situation in which labour demand exceeds supply, resulting in rising wages. This situation is sometimes described as the 'neo-classical model' of international migration, since it suggests that the labour market is self-regulating through the wage mechanism, with unevennesses in international wage levels stimulating migration flows which bring the supply and demand for labour once more into balance.

Detailed study of the history of labour migration in the Arab region shows that flows have not been random in their patterning nor have they reflected the operation of some equitable and disinterested market force. It is useful first to investigate the pattern of labour migration to show how, through time, the source of immigrant labour has changed as a result of the specific preferences and actions of the oil-rich states, and second to study the relationship between net migration flows and trends in other economic indicators.

Given the pre-existing history of labour migration between Arab countries, the shared cultural and linguistic ties, the weak levels of economic growth being experienced in the early 1970s by the Arab states without oil, and the rising levels of unemployment in these states (Jordan for example had an estimated 14 per cent of its workforce unemployed in 1972), it was not surprising that the 1973–4 oil shock was followed by an in-flow of large numbers of immigrant workers to the oil states. By 1975 there were over half a million Jordanians and Palestinians working in Saudi Arabia. In addition over 300,000 Yemenis and 200,000 Egyptians had been absorbed into the Kingdom's rapidly expanding labour force. Libya had become home to over 300,000 Egyptians, while Kuwait had over 200,000 Palestinians and Jordanians. Some countries were more willing than others to accept migrant families. This is shown in Table 5.2 in terms of differential activity rates. It can be seen that while most migrants to Saudi were actively employed in the labour force and were not accompanied by their wives and children, in the smaller sheikdoms of the Gulf migrants, and in particular Palestinians, often migrated with their families. This had the effect of lowering migrant activity rates. It increased the cost of supporting the migrant community for the host nations and reduced the return flow of remittances to the countries of migrant origin.

Too much emphasis should not be put on the statistics quoted above since estimates of migrant population sizes have often been disputed between sender and receiving countries. The main reason for listing these figures is to show that migrant flows were not random but geographically ordered. Most labour-sending countries sent the majority of their migrant workers to one, or at most two, of the oil-rich states. There are good reasons within established migration theory which explain why the process in itself usually has an in-built bias which favours flows forming geographically concentrated patterns. From a development perspective, the important feature is that a geographical dependence emerged between labour-sending and labour-receiving countries. While labour receivers usually drew on a wide range of sources, labour senders were often highly dependent on one foreign labour market. This amongst other features meant an inequity of power between labour supply and labour demand.

The inequity of power over labour market opportunities in the Arab oil states was expressed in different ways at different times. Occasionally it was used by political leaders such as Colonel Gaddafi to reinforce other aspects of international policies. For example on several occasions in the 1970s he expelled both Tunisian and Egyptian migrants from Libya in attempts to de-stabilize his neighbours by flooding their labour markets with return migrants. Of more significance the inequity of labour market power was expressed through the regulation of migrant wages. As the direct costs (wages) and indirect costs (support of immigrant families by housing, health and education services) of immigrant workers from the Arab countries began to rise, new sources of labour were increasingly tapped. In particular Asian labour was found to be both cheaper and more pliable than Arab labour. From the mid-1970s through the 1980s Asian labour was increasingly introduced. In addition to Pakistan and India, Bangladesh, Sri Lanka, the Philippines, Turkey and Thailand all became significant senders of labour to the oil states. By 1988 it was estimated that there were 400,000 immigrant workers from Pakistan, 300,000 from the Philippines and 145,000 from Thailand in Saudi Arabia, while by 1990 the Asian share of the migrant workforce of the Gulf Cooperation Council states was believed to have risen to 55 per cent of the total. This switch from Arab to Asian labour is also shown by statistics for migrant remittances, which for most Arab labour-sending countries peaked in 1983 and 1984 before dropping in the latter half of the 1980s.

Although in theory most of the Arab oil states endorsed migrant recruitment policies favouring the introduction of Arabs over other migrants, in practice this did not happen. The history of labour migration to the Arab oil states demonstrates the geographical dependence of the Arab labour-supplying countries upon the oil-rich states and their inability to prevent a switch from their labour supply in favour of other sources through time. In the aftermath of the 1990–1 conflict in the Gulf, Asian and Egyptian labour has replaced migrant stocks from Jordan, Palestine and Yemen, reflecting Kuwait's and Saudi Arabia's distrust of those migrant communities which supported the Iraqi cause. Jordan was hardest hit with the United Nations Relief and Work Agency estimating that about 300,000 migrants (Palestinian and Jordanian) returned from Kuwait and other Gulf states after the war. They faced an existing unemployment level in Jordan of around 20 per cent. Some estimate that the subsequent temporary collapse of certain sectors of the Jordanian economy saw unemployment soar to over 30 per cent.

A second theme suggested in the introduction to this chapter is the relationship between the volume of net migration flows and indicators of national economic growth in the oil states. If, as some have hypothesized, volumes of international labour migration can be interpreted as the balancing mechanism bringing labour demand and supply into equilibrium through the market mechanism of differential wage levels, then one might anticipate that rising oil revenues and positive economic growth would be associated with rising migration stocks in the oil states, while falling oil prices would act as a catalyst to net return migration and declining migrant stocks. This relationship will be described from now on as the 'neo-classical economic view'. The term migrant stock will be used to refer to the total size of the migrant population in a country, in contrast with the term 'migrant flow' which is used to refer to the number of immigrants or emigrants moving between states over a limited time period (usually one year).

Birks *et al.* (1986) discovered from comparison of time series statistics that parallel trends existed for the years 1975–84 between work permits issued to migrants and levels of government expenditure (strongly related to oil revenues) in the oil states. For example, in Oman they found an almost perfect correlation; in Abu Dhabi government expenditure accounted statistically for about 82 per cent of the variation in the number of migrant permits; while in Kuwait about 56 per cent of the statistical variation was explained. It is seldom, in international migration analysis, that such strong statistical explanation is possible in terms of a single determining variable. On this basis they anticipated that 'retrenchment of economic activity in the Gulf should have the same outcome – a reduction of demand for migrant labour and the re-export of "surplus" immigrant workers' (Birks *et al.*, 1986, 801).

The slump in world oil prices was widely predicted to reverse the trends described above and to stimulate a considerable exodus of labour migrants. Research shows, however, that the neo-classical economic view does not easily match the experience of Arab labour markets in the middle to late 1980s and that for a variety of reasons foreign labour was retained by the oil economies.

Several different aspects of Arab labour migration in the 1980s need to be explained. First there were significant return migration flows from the oil states long before the main crash in oil prices took place. During the first half of 1985, labour migrants were reported to be leaving Saudi Arabia at the rate of 60,000 per month. This trend was mirrored in other oil states and can be explained more easily by the new phase of development which had been reached prior to the economic recession. While in the 1970s Saudi Arabia and the other oil states were engaged in major infrastructural development projects, involving substantial labour forces for construction-related activities, by the mid-1980s this phase of development was coming to an end and Saudi Arabia had to face a situation in which opportunities for productive infrastructural investment were virtually saturated. The demand for migrant workers switched to personnel able to service and maintain the projects initiated a decade earlier.

The 1986 crash in oil prices did of course have some effect on labour demand. In the first half of 1986 prices fell from US$26 to US$8 a barrel (i.e. to 36 per cent of the former value). The statistics for migrant stocks in Saudi Arabia show some decline, but instead of being cut to a third of its former size in line with the scale of the decline in oil prices, there was a reduction of only about 10 per cent. Statistical evidence suggests, therefore, that volumes of return migration were not particularly strongly affected by fluctuations in world oil prices. It would appear that once dependent on migrants to carry out certain types of activity the oil economies found it hard to reverse this situation.

Although there was no mass exodus of migrants, the crisis did provide a political context for the introduction of restrictive migration policies. For example, Saudi Arabia in its Fourth National Plan (1985–90) used this as a pretext to 'make room' for its own population in certain sectors of the labour market and announced the objective of reducing its foreign workforce by 600,000 persons. In 1986 it introduced legislation to restrict migrants changing from one employer to another. The United Arab Emirates announced plans to 'nationalize' all 50,000 government jobs, while 'Omanization' became a major policy objective of the Sultanate of Oman. The result of these policies was to create fear amongst the migrant community of repatriation and to strengthen the power of employers with regard to wage bargaining. Wage levels were reported as dropping by as much as 40 per cent in some Gulf countries in the mid-1980s, while the increasingly restrictive policies also favoured the growth, as in Western Europe in the 1970s, of clandestine migration. One estimate suggests that the number of clandestine migrants in Saudi Arabia rose from around 150,000 in 1984 to 230,000 in 1987. The primary effects of the slump in oil prices in the mid-1980s was therefore a reduction in migrant wage levels and an increase in clandestine migration. It was not a vast reduction in migrant stocks.

The imposition of overtly restrictive policies on immigration changed the composition and character of migration, as well as significantly modifying the conditions of employment. In place of the neo-classical view of migration, analysis of the relation between economic forces and migration trends in the oil states has therefore produced a rather different picture. If the late 1970s saw a shift in demand for migrant labour from dominantly Arab to Asian sources, the mid-1980s saw a further shift in demand, which had the effect of consolidating migrant stocks and reducing wage levels. These two changes in demand for migrant labour were intimately linked in a historical sense with economic trends, but rates of economic growth and decline are not statistically correlated over the decade in a linear fashion with trends in the size of migrant stocks. Shifts in demand saw Arab migrant labour first of all forced into increasing competition with labour from Asian sources and then forced to accept much lower wages under conditions more strictly controlled and policed by state authorities.

Housing and labour market implications of migration

The development implications of labour emigration vary from country to country depending on the nature of their labour forces and the way that emigration interacts with the character of particular economies and societies. Case study G relates to the experience of the Yemen Arab Republic. It is important, however, to search for broader evaluations of the consequences of labour emigration within the Arab sender countries, since it has been a common and important feature of so many of the non-oil-producing Arab states during the last two decades.

Case study G

Yemen: the world's most migrant-dependent state

The Yemen Arab Republic was unusual in the Arab world in having received very little attention from the colonial powers of Western Europe. This largely reflected the Western perception of the region as not having any significant mineral, agricultural or strategic importance. The result was that even in the middle of

---

Case study G (*continued*)

the twentieth century North Yemen, as it was often described, remained an Islamic *imamate*. That is, it was still governed by *imams* or religious leaders, with the primary level of social organization remaining the tribal unit. Between 1962 and 1970 there was a period of civil war which devastated what little infrastructure there was. When the war ended the Yemen Arab Republic epitomized in almost every respect the model of an under-developed economy.

The country's first census in 1975 revealed that 47 per cent of the population was under 15 years of age, 78 per cent of employment was in the agricultural sector, 83 per cent of the population was illiterate, 93 per cent of the population lived in rural areas, there were 24,000 persons for every doctor and life expectancy at birth was only 38.5 years. The country also had a severe inability to absorb modern technology owing to widespread skill shortages and the institutional weakness of the main state organizations responsible for guiding and implementing development plans. In addition, the widespread consumption of *qat*, a mild narcotic, severely lowered labour productivity in all sectors of the economy.

Plate 5.1 Craftsmen chewing *qat* in the jewellery *suq*, Sanaa

Case study G (*continued*)

Plate 5.2 Fortress house, western highlands, Yemen

It was in this context of considerable poverty, combined with a poor capacity for self-development, that the Yemen Arab Republic was suddenly confronted with new opportunities for contact with the world economy in the form of labour migration to the oil-rich states. Its proximity to Saudi Arabia ensured that in the early 1970s hundreds of thousands of Yemenis departed to work abroad. Given the nature of Yemeni under-development it is not surprising that Yemenis formed one of the least skilled groups of migrants in Saudi Arabia, being involved largely in manual employment in the construction and agricultural sectors. For many of the Yemeni

Case study G (*continued*)

migrants the move was not so much a move from a low wage to a high wage economy as an escape from subsistence agricultural employment in the Yemen Arab Republic under what might be thought of in European terms as feudal conditions. Despite the relatively low wages received by the migrants while abroad, the scale of the migration flow was such as to result in a significant inflow of remittances to the Yemen Arab Republic. By the 1980s remittances were valued as worth the equivalent of 30 per cent of the country's GDP and they were valued at several times the earnings of the country's exports. Indeed in the early 1980s the ratio of remittances to the value of commodity exports was higher than in any other country in the world. In a very short period of time international migration became the main motor driving the economic development or under-development of Yemen.

What were the development implications of the Yemen Arab Republic's heavy dependence on labour migration? The effect on the labour market was particularly dramatic. The withdrawal of a large part of the rural labour force to work abroad resulted in rural wages rising rapidly. Between 1975 and 1980 rural wages experienced a 500 per cent increase in value. In the absence of between a third and a half of adult males from some villages during the peak years of emigration in the late 1970s, it is not so surprising that women and older children were increasingly drawn into the rural labour force (Findlay, 1987). Despite rising rural wages emigration remained a very attractive option to young landless Yemenis. It has been estimated that in the late 1970s they could earn six to seven times as much by working in Saudi Arabia as by staying in rural Yemen. Emigration of adult labour from the towns was also significant. Boys found it easy to obtain jobs left vacant by their elders who had emigrated. This undermined the government's efforts to encourage higher educational standards since young people could see no benefit in labour market terms in continuing with their education.

In the countryside, land use changes occurred very rapidly as production patterns adapted to the escalating cost of labour and changing demands for agricultural produce. In order to achieve adequate returns on labour, low value crops and land uses were

Case study G (*continued*)

abandoned and the rural workforce focused its efforts on producing cash crops. As a result, large areas of terraced agriculture geared towards the production of subsistence grains (such as sorghum, wheat and millet) went out of production. The 1983 agricultural census showed that 16.6 per cent of the Yemen Arab Republic's cultivable land lay abandoned. Effort in the agricultural sector was switched to high value crops (fruit, vegetable and *qat*) and to the rearing of poultry. In particular the area under *qat* production expanded, since the accumulation of domestic capital from remittances permitted higher levels of *qat* consumption.

The in-flow of migrant remittances to the Yemen Arab Republic highlights its difficulty in profiting from international migration. The Central Bank of Yemen was only established in 1971 and even in the 1980s the distribution of local commercial banks was sparse. As a result a considerable proportion of remittances were transmitted by private remittance agents. These agents arranged migrant visas, looked after remittance transfers and organized migrant savings in their region of origin. Once money had been sent home it was often used to finance the construction of a new house or the purchase of consumer goods, such as radios, kitchen electrical items or perhaps a small van. One of the results of this new pattern of expenditure was to boost the level of foreign imports to the Yemen markedly since even in the early 1990s it had a minimal level of manufacturing production. The net effect of migration on Yemen's balance of trade was therefore far from desirable. Although remittances were a welcome source of foreign income, a large proportion did not flow through official banking channels. Instead of financing significant domestic investment they served instead to boost foreign imports, at the same time as changes in the agricultural economy were necessitating a rise in food imports to replace the decline in the subsistence production of basic food crops.

Remittance receipts record something of the traumatic history of Yemeni emigration. Remittance levels rose rapidly to a peak in 1978–9 and then tailed off as Yemeni labour stocks in Saudi Arabia stabilized and as Yemeni migrant wage levels began to drop as a result of the influx of cheaper unskilled Asian labour.

Case study G (*continued*)

The imbalance of trade which resulted from the Yemen Arab Republic's soaring import bill led to the Yemeni riyal being devalued to approximately half its former value in the mid-1980s. One result of this was that many migrants hoarded their savings in foreign currencies to the disadvantage of the Yemeni economy. The switch in labour market circumstances in Saudi Arabia in the mid-1980s saw wage levels of all migrants fall dramatically and for some Yemenis resulted in unemployment and the need to return home. This caused a further drop in the country's remittance earnings (worth only US$242 million in 1989 compared with US$1,084 million in 1983) and the onset of a severe financial crisis. The expulsion of nearly 1 million Yemenis by the Saudis in 1990–1 brought the final blow to this aspect of Yemeni development history. The rural labour market of one of the world's least developed countries was flooded by a mass involuntary return migration.

The oil states were eager to engage migrants on a short-term basis in order to help achieve their development plans, but with no intention of allowing the migrants to settle and integrate in their societies. Most migrants were employed initially on 'batchelor status' contracts of one or two years, after which time they were expected to return home. In practice many Arab migrants renewed their contracts, thus extending their residence in Kuwait, Saudia Arabia and the other Gulf states. Temporary migration was gradually transformed into a rather more permanent feature. One dimension of the host country's immigration policy was to favour the spatial segregation of immigrants from the indigenous population. This was particularly marked with regard to Asian immigrants and Western expatriates. Asian migrant groups were often located in work camps on former desert sites. These planned enclave developments minimized contact between immigrants and the indigenous populations. Examples included work camps established at Ruwais (Abu Dhabi), Jebel Ali (Dubai), Shuaibi (Kuwait), Umm Said (Qatar) and Jubail (Saudi Arabia).

With a few exceptions, immigrants failed to gain the citizenship of the host countries. They were often joined by their families, however, thus changing their housing and service needs. In Kuwait the number of houses built for foreigners increased from 17,000 in 1957 to over 100,000 in 1980. And by the 1980s providing education for immigrant children accounted for approximately 50 per cent of the Kuwaiti education budget. Inevitably much of the migrant housing had to be built very rapidly. It was in part financed by the native populations of the Gulf states who found renting property to immigrants a lucrative new source of income. Migrants wanting to minimize their expenditure on rents and to maximize their savings often accepted overcrowded and low standard accommodation. The result of these interacting forces was the rapid growth of a highly differentiated urban landscape in the oil states, with contrasting residential areas for indigenous and immigrant populations. Bourgey (1984, 233) graphically describes the situation in Abu Dhabi, whose population grew from 47,000 in 1968 to 451,000 in 1980 mainly as a result of the massive influx of immigrants:

> Within several hundred metres of the most luxurious hotels of Abu Dhabi, behind a road with some of the most beautiful shops in the UAE and right in the very centre of the capital, has developed an immense shanty town of Pakistanis. In the Emirates the urban landscape expresses on all sides the contrasts and inequalities of urban society.

A sample survey of migrants in Kuwait undertaken by Al-Moosa and McLachlan (1985) showed that 72 per cent of Palestinian and Jordanian migrants had been in Kuwait for fifteen years or over while only 16 per cent of Egyptian migrants had been resident for this length of time. This contrast highlights several important features of Arab labour migration. First, political as well as economic factors have been important in accounting for the evolution of the migrant community, with refugees forming a significant element of the migrant stock. Second, through time some migrant groups seek to settle in the oil states, even though they are not being granted citizenship and despite the initial intentions of the host societies to permit only temporary labour immigration. Associated with the increasing permanence of some migrant groups has been the tendency for these groups to seek increasingly to be re-united with their families through the immigration of their dependents. This, as noted above (Table 5.2), reduced immigrant activity rates and pushed up the cost to the host society of using immigrant labour through the need to provide more housing and urban services (health, education) for their families.

For the labour-sending countries the effects of emigration on their economies change through time, as a result of the increasing permanence of what started as a temporary flow. While income from remittances continued to be a welcome source of foreign currency earnings, the size of remittances sent back per capita tended to decrease as migrants were joined by their families. In addition the shift to lower migrant wages had an adverse effect on the remittance levels received by several of the Arab labour-sending countries in the late 1980s (Table 5.1). Over time the negative effects on the labour markets of the sender countries have become more evident. The case study on Yemen has shown how the withdrawal of a large part of the rural labour force had the effect of pushing up wages and reducing agricultural production of key cereal crops. In Jordan and Egypt similar labour shortages were experienced. In Jordan agricultural production was maintained by encouraging Egyptian immigration, while Egypt itself received Sudanese immigrants. By 1985 secondary labour immigration to Yemen meant that remittances leaving the country had risen to 7.6 per cent of the value of remittances received from Yemenis in Saudi Arabia, while in the same year 23.1 per cent of Jordan's remittance receipts were offset by remittances leaving the country. Labour-sending countries therefore also became labour receivers. The migrants entering these countries often accepted lower wages than native labour, making it unlikely that return migrants (e.g. Jordanians returning to rural Jordan) would accept agricultural employment once more and making re-insertion of return migrants from the oil states problematic.

Some have argued that one of the great benefits for labour-sending countries is that migrants may return with new and useful skills. In the countries of the Maghreb it was certainly true that some migrants working in France were employed in very different sectors of the economy from those in their countries of origin (e.g. Tunisians working in the car assembly plants of Renault), but return migrants seldom found the opportunity to use newly acquired industrial skills on their return home. Many Jordanians were already highly trained before emigrating, as shown in a survey carried out by the author in 1987 amongst return migrants (Findlay, 1987). Most worked in the same sector of the economy while abroad and acquired very few new skills during their spell as migrant workers. In the Yemen Arab Republic by contrast, most emigrants had been drawn from rural origins. Some experienced urban employment while in Saudi Arabia and returned with useful skills such as in the construction sector. A survey of over

2,600 building companies in Yemen by Meyer (1986) showed that no less than a third of all employees had gained their apprenticeship in the building trade while abroad and had returned home to apply their new skills in activities such as the manufacture of wooden doors, iron gates and water tanks and the production of tiles and concrete blocks. The net benefit of emigration for the labour-sending countries in terms of skill acquisition must therefore be seen as varying between one country and another. It would seem that the benefits were greatest to Yemen because of that country's low level of labour skills and because of the nature of the training which some emigrants received while abroad.

The use of remittances

Studies of the effect of migrants sending home part of their wages to their country of origin provide interesting information about the potential development effects of international migration. Most studies point to a small reduction in the level of poverty in the countries of origin as a result of the in-flow of remittances. For example, a study of Moroccan emigrants showed that migrant families were supported at a higher standard of living than would otherwise have been the case. There was, however, the negative effect that some migrant families gave up agricultural production because they found that they could support themselves on remittances, which had the long-term effect of increasing dependence on income from foreign labour markets. In Egypt a study by Adams in 1991 showed that, while remittances reduced the overall level of poverty, they increased income inequalities, a finding repeated in many other labour-sending countries.

Analysis of remittance use shows that most migrants do not invest in long-term agricultural or industrial projects. Instead, by far the most common destination for remittances is investment in housing, housing repairs or land for building. Adams (1991) found this in 54 per cent of cases. Table 5.3 shows similar results from a survey of 250 migrant households in Jordan. Some 40 per cent of return migrants had also used part of their foreign income to purchase a car, while about a third had invested in land. In the early 1980s investment in land close to the capital city proved a very effective way of increasing the value of remittances since land values were rising rapidly during a phase of speculative urban growth. Less than 8 per cent of remittances were used in productive agricultural or industrial projects. This and other research show that, while remittances reduce the effects of poverty and lead to positive improvements in housing conditions, they do not have as great an effect on the long-term development of migrant-sending countries as one might hope. Remittance expenditure patterns are dispersed and relate to the individual ambitions of the migrants and their families rather than to the larger-scale type of investments which could generate more sustained long-term economic development.

Table 5.3 Remittance use by return and current migrants in Jordan, 1984

Remittance use	Return migrants (%)	Current migrants (%)
Building (new or repair)	55	43
Car purchase	40	20
Land	31	21
Education	17	20
Industrial project	5	4
Agricultural project	3	1

Source: Findlay and Samha (1986, 176)
Note: Proportions are expressed as the percentage of persons investing in each category.

International migration and settlement patterns

One of the most interesting effects of international migration in the Arab world has been on settlement patterns in the regions of origin. In some cases, such as rural Oman, the collapse of the rural economy followed the removal of key elements of the labour force through emigration, the result being rural de-population. In other areas, as in the Rif mountains of northern Morocco, parts of rural Egypt and in the villages of northern Jordan, the result has been a new source of rural income and the possibility of sustaining rural populations which would otherwise have been likely to migrate internally within these countries in search of urban employment (Seccombe and Findlay, 1989). In poorly serviced rural areas international remittances may have the long-term effect of favouring rural–urban migration because of the desire of migrants to build housing in locations serviced with a good infrastructure and, in particular, access to electricity. Another force favouring urbanization has been the desire by return migrants in contexts such as the Yemen Arab Republic to use newly acquired urban skills. This then favours the transfer of migrants who originated in rural areas to urban locations on their return.

Case study H

Migration and the urban growth of Amman city

Until the twentieth century Amman was only a small village sited at the junction of several of the wadis which traverse the dip slope of the limestone plateau to the east of the Jordan rift valley. The decision by the Ottomans to use the wadi when they constructed a railway from Damascus to the city of Medina in the Hejaz increased the importance of the location slightly. Amman was still a small town of only 5,000 inhabitants when it was chosen as capital of the newly created Hashemite Kingdom of Transjordan in 1922. By 1990 it had grown to a conurbation of about a million people and a surface area of over 50 square kilometres (Figure 5.1).

The demographic and spatial expansion of Amman is due more than most cities in the Arab world to refugee and international migration. Amman's administrative functions did little to fuel its

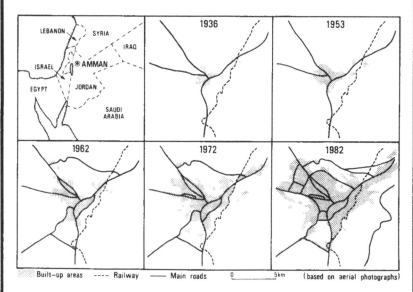

Figure 5.1 Amman's urban growth
Source: Findlay and Samha (1986, 180)

Case study H (*continued*)

Plate 5.3 Speculative urban development on the outskirts of Amman, Jordan

early growth, since it was established as capital of a new state with few modern roads, little by way of urban infrastructure and a surrounding state where only 3 per cent of the surface area was cultivated. The first significant growth occurred in 1948–9 when some of the refugee flows, stimulated by the creation of the state of Israel, focused on the city and resulted in the creation of a ring of so-called camps. In practice, the grid iron pattern of these areas of refugee settlement continues to be a distinctive part of the morphology of modern Amman. From 1949 to 1967 internal migration from other parts of Jordan and from the West Bank produced urban sprawl and the establishment of some low income suburbs. Then in 1967 the Israeli occupation of the West Bank created a new wave of refugees to Amman and a further circle of refugee camps around the city. The city's economy by now had been transformed by the in-flow of aid to support the refugee population and by assistance from other Arab states to help Jordan in its role as a front line state with Israel. There was insufficient employment, however, to absorb the refugee population and so, not surprisingly, many Palestinians subsequently

Case study H (*continued*)

departed to work in the oil states, leaving their families behind in the safe havens around Amman city. As with the earlier camps, those created after 1967 soon became permanent features. By the 1980s many of the inhabitants of the camps were children born to the original refugees, but with no more desire or ability to become permanent residents of Jordan than their parents.

During the 1970s and 1980s the pace of population growth in Amman slackened with no more major refugee waves. Spatial expansion, however, continued at a rapid pace with migrant families using remittances to improve their housing and to build new apartment blocks and luxury villas, many for letting or speculative purposes. This created the curious paradox of a city encircled with overcrowded refugee camps and areas of relatively poor housing occupied by families who had no relatives working in the oil economies (the Amman planning authorities estimated that 80 per cent of households in Amman in the 1980s had such low incomes that they could not afford even the minimum desirable housing) and at the same time the existence of thousands of empty and unlet luxury apartment blocks.

Amman city is one location where the effect of the international migration process on urbanization is particularly obvious (see Case study H). Many Jordanian and Palestinian migrants chose to use their remittances to invest in housing in Amman. This housing has been constructed not altogether for the direct benefit of migrant families but as a speculative investment in real estate. Most of the new housing has abandoned the courtyard plan of the traditional Arab home and has adopted the extroverted style of European housing with a central residential unit and surrounding garden. Surveys have also shown other ways in which migration has affected urban forms and lifestyles. For example, return migrants had a higher level of ownership of electrical items and imported household consumer goods, and there seems to be some evidence to suggest that the migration experience has led to families accepting a more 'Western consumerist' lifestyle. Some return migrants carry with them ideas of urban innovations alien to the urban cultures of their areas of origin. For example, in traditional Arab cities like Sanaa in Yemen it was return migrants who were responsible for setting up shops in the form of Western style supermarkets and for introducing fast food outlets and Western restaurants. In Amman emigration has affected other aspects of urban society as in the increase in female participation in shopping for clothes and food (Findlay, 1990). This occurred because of the transfer of household roles to the wife during the absence of the migrant as head of household.

Changing times; changing migrants

This chapter has explored some of the diverse development implications arising from labour migration within the Arab world. It has been suggested that Arab labour migration trends in the 1980s and 1990s cannot be explained in terms of international migration theories which rest upon the assumptions of market forces producing a balance of labour demand and supply through the migration mechanism. International labour market development, as with other aspects of economic development, has involved a struggle between unequal economic forces. Migrants selling their labour in the oil-rich states have found that political, economic and military circumstances have permitted the labour-importing countries to assume a dominant role in determining which Arab and non-Arab countries send most migrant workers and at what wage level. This was illustrated during the 1980s by the switch away from Arab to other Asian labour market sources (in Saudi Arabia and the United Arab Emirates Arab immigrants declined from 71 per cent to 42 per cent between 1975 and 1985) and by the strong discontinuity in the composition of immigrant labour stocks before and after the Gulf conflict of 1990–1.

War has been a major stimulus to changes in labour migration flows on several occasions in the 1980s and 1990s. Little studied, yet numerically of vast significance, was Egyptian labour migration to Iraq during the 1980s. A large proportion of Iraqi males were withdrawn from the labour force to serve in the army during the eight-year war with Iran. Their civilian jobs were mainly filled by Egyptians. At the end of the Iran–Iraq war many Egyptians found that their services were no longer needed, but up to half a million remained in Iraq until the 1990 invasion of Kuwait when they too had to return home. Table 5.4 shows that they were the single largest migrant group affected by the 1990 war, although the plight of the million foreigners working in Kuwait at the time of the invasion received more attention from the Western media. Other major return flows stimulated by the war included the Yemeni workforce of Saudi Arabia. Following the war Jordanians and Yemenis found it hard to gain new contracts to work in Kuwait and Saudi Arabia, while Egypt found its loss of access to the Iraqi labour market partially offset by the vacancies arising in Saudi Arabia. Asian labour suppliers were eager to take up positions left vacant by Jordanians and Yemenis, with the government of Sri Lanka even offering to pay for the transportation of migrant workers in order to secure jobs in the new labour market situation (Meyer, 1992).

Table 5.4 Estimated numbers of selected migrant groups in Iraq and Kuwait, July 1990

Country of origin	Kuwait	Iraq
Jordan/Palestine	380,000	17,000
Egypt	170,000	500,000
India	162,000	8,000
Sri Lanka	92,000	n.a.
Pakistan	90,000	10,000
Bangladesh	70,000	15,000

Source: adapted from Meyer (1992, 119)
Note: n.a., not available.

Despite the problems which labour emigration has brought to most of the Arab sending countries over the last two decades, it would be wrong to conclude without once again stressing that it also introduced certain opportunities. For Yemen, Jordan and Egypt, emigrants added significantly to the foreign earnings of their countries of origin. In terms of international comparisons these economies were in the 1980s amongst the most dependent on migrant remittances in the world. So great was the dependence that it produced distinctive economic and social features in the development paths followed by these states. This is an issue raised again in Chapter 7 in relation to the urban economies of Arab countries. From the perspective of this chapter what is important is that the labour migration flows of the last two decades produced a specific form of economic interdependence in the Arab region. Although not without its benefits, dependence on emigrant remittances has been shown to be a highly vulnerable path to economic development.

Key ideas

1 In the 1980s Yemen, Jordan and Egypt depended on income from migrant remittances to a greater extent than almost any other states in the world. In these countries labour emigration proved a key characteristic of the development of their economies.
2 The slump in world oil prices in the mid-1980s did not trigger a mass exodus of migrant workers but resulted in downward pressure on migrant wages.
3 International migration has acted as a catalyst to economic and social change in countries of origin, but the changes have not always been either desired or desirable. Remittances have been spent to a large extent on consumer goods or improved housing, with few migrants investing in agricultural or industrial projects.

6
Rural development

Introduction

The rural landscapes of the Arab world are as diverse and complex as those of any other major culture region. To generalize about the activities of pastoralists and cultivators in such a vast area is as difficult as to describe and explain in general terms rural economic activities in regions as diverse as Western Europe. Consider, for example, the vast contrasts of ecological, economic and social circumstances which have produced the areas of sugar beet production in the Gharb plain, the plantation agriculture of the vast orderly olive groves of Sfax, the date palms of the Djerid oases, the irrigated vegetable and cereal production of the Fayoum Depression, the intricate terraced agriculture of the mountains of Yemen, the vegetable gardens of Sanaa, the great circles of grain production that dot parts of Saudi Arabia, the extensive nomadic pastoral movements of the Ruwala tribes, or the intensive cloche cultivation systems of the Jordan valley. Clearly, this chapter can only touch on a very limited number of the more important issues affecting rural development.

How can rural development be assessed? Rather than comparing the experience of particular agricultural economies with some abstract western model it seems more profitable to commence by analysing what has been achieved in a particular locality relative to the resources available in that place. Accordingly this chapter commences by considering the most critical resource for all rural dwellers in the Middle

Plate 6.1 Irrigation canal and agriculture, Fayoum Depression, Egypt

Plate 6.2 Olive plantations, the Sahel region of Tunisia

Plate 6.3 Threshing floor by a dry wadi in the eastern highlands of Yemen

East and North Africa: water. The Sahara and Arabian deserts with their bordering semi-arid zones are the most extensive hot arid region in the world. Although human responses to the shortage of water have varied greatly across the Arab countries, rural dwellers are united by the common challenge of surviving in a harsh and often unpredictable marginal environment. This chapter considers ways in which water resources have been used and adapted in different rural settings to provide a livelihood for the populations of the Arab countries. Nomadism is a pastoral activity widely perceived to be of importance to the region since it is one way in which rural societies have adapted to arid environments. Much more important as a source of employment and of food are the practices of crop cultivation which exist. The chapter concludes by considering ways in which governments have sought to intervene for ideological and economic reasons in the food production systems of their states.

Water: a critical resource

Although a moisture deficit (a situation in which potential evaporation exceeds precipitation) exists across most of the Arab world, water can be found in some of the most arid areas. This occurs either where rivers entering this moisture-deficient region transfer surplus water from surrounding lands into the arid zone or where groundwater reserves exist, reflecting storage of water in underground reservoirs charged by former pluvial phases or by long-distance aquifers. Almost without exception where water resources have been discovered it has led to the peopling of the associated rural areas of the Arab world. Where water resources do not exist, life becomes almost impossible to support for any length of time because of the extremely high temperatures (a July average of over 30°C) found along the tropic of Cancer. Across much of the Arab world it is meaningless to quote annual precipitation figures since rainfall may occur once in twenty or perhaps fifty years. Precipitation in these environments is therefore rare and unpredictable, but when it does fall in a particular part of the desert it may come in the form of a heavy rainstorm causing temporary flooding.

A population density map of the Arab countries continues to reflect remarkably closely the pattern of water availability. Higher population densities are found in areas such as the mountains of Yemen, Lebanon, Morocco, Algeria and Tunisia where reliable seasonal precipitation recharges groundwater levels; in the Nile, Jordan, Euphrates and Tigris valleys where river regimes have for millennia permitted intensive agricultural activity; and in the array of oases which are sprinkled across the sparsely populated areas of the Arab world.

Not only is most of the region very arid, but those areas which do receive modest amounts of rainfall experience a strong seasonal concentration of precipitation. This produces many temporary river courses or wadis which flow for a few days or weeks but are dry riverbeds for the rest of the year. Wadis do, however, serve to concentrate rainfall, creating lines across the landscape where both short-term and longer-term recharging takes place and where consequently the grazing of animals is possible over at least part of the year. Emanating from the mountains of the region one therefore finds interesting patterns of rural land use, with fingers of better land for grazing often extending out of the mountains far into very arid areas. Such lines exist because of seasonal discharge along wadis or shallow aquifers. For example these patterns are evident along the eastern side of the Yemeni highlands on the margins of the Rub al Khali (Empty Quarter) and along the southern fringes of the High Atlas. Human societies have of course copied natural processes by building water channels from the mountains out onto surrounding plains. One interesting aspect of this are *falaj* (known as *qanat* in Iran) or underground channels built by rural societies in several parts of the Arab region (e.g. Libya and Oman). The underground passages minimize evaporation of the water during its transfer. *Falaj* depend on gravity to sustain water flow and so require careful maintenance to clear sediment if water flow is to be sustained. When properly maintained, *falaj* permit very intensive oasis style agriculture to be located in otherwise extremely arid areas.

Most rivers in the Arab world are seasonal because they depend on local catchment areas in which precipitation falls over only a few weeks of the year. If it was not for the size and diversity of its catchment basin even the Nile would dry up. Its water is drawn from two enormous catchment areas that cover about 10 per cent of Africa. To the east the Blue Nile taps the late summer monsoonal rains of the Ethiopian Highlands and contributes about 80 per cent of the total Nile flow at Aswan, while to the south the White Nile's tributaries extend to spring rain catchment areas around Lake Victoria. Before the dam was built across the Nile at Aswan in southern Egypt, peak discharge was about 800 million cubic metres per second during the months of August to October, but fell to about 100 million cubic metres per second during the rest of the year. This unusual pattern has been described by the fourteenth-century Arab traveller Ibn Battuta:

> The Egyptian Nile surpasses all rivers of the earth in sweetness of taste, length of course, and utility. No other river in the world can show such a continuous series of towns and villages along its banks or a basin so intensely cultivated. . . . One extraordinary thing about it is that it begins to rise in the extreme hot weather, at the same time when rivers generally diminish and dry up, and begins to subside just when rivers begin to increase and overflow. . . . The inhabitants of every township have canals led off the Nile; these are filled when the river is in flood and carry water over the fields.
>
> (Ibn Battuta, 1929, 52)

The significance of the floods was not merely that they brought water to the carefully cultivated fields of the Nile valley and delta, but that they also brought rich sediments. It has been estimated that while in flood the Nile spreads some 50 million tonnes of silt over the land.

Before turning to consider in more detail human intervention in the hydrological cycle it is interesting to note the current discharge situation

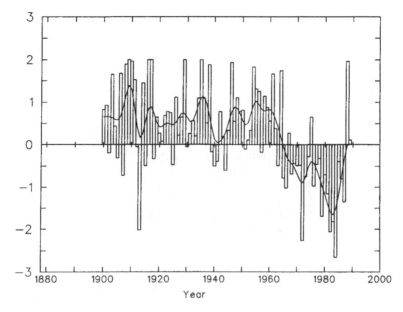

Figure 6.1 Deviation in annual discharge for the Blue Nile at Khartoum (metres), 1900–89
Source: M. Hulme, CRU, University of East Anglia

of the Nile. This has been affected not just by dams, canals and irrigation schemes, but also by long-run variations in the precipitation patterns in its catchment. Figure 6.1 shows the standardized annual discharge (DAI) of the Blue Nile at Khartoum, while Figure 6.2 records the deviations in the lake level of Lake Victoria. On both graphs the continuous line is a ten-year filter used to smooth the time series information and to present an expression of overall trends in discharge.

The graph for the Blue Nile shows that for the twenty-five years from the mid-1960s to the late 1980s discharge was consistently below the river's 'normal' level. This reflected two important features. First there was the now well known sequence of years of very low or zero rainfall which produced sustained drought in Ethiopia. Second there was the effect of water abstraction from the Blue Nile south of Khartoum to supply the Gezira/Maraqil irrigation scheme. Fortunately for Sudan and Egypt the reduced discharges of the Blue Nile have been offset by a sequence of wetter years in the catchment area around Lake Victoria, as reflected in the higher lake levels of the 1960s and 1970s. For Egypt,

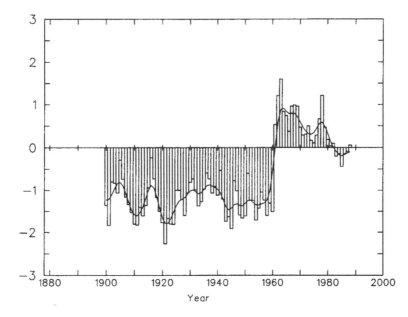

Figure 6.2 Annual lake level deviation for Lake Victoria (metres), 1900–88
Source: M. Hulme, CRU, University of East Anglia

water stored in Lake Nasser was crucial in preventing a severe water shortage in that country in the late 1980s.

Dams and irrigation

Given the strong seasonality of rainfall and river discharge in the Middle East, it is not surprising that irrigation systems and dams have been used since earliest times in an attempt to store and redistribute flood waters. One dramatic example is the great dam built by Queen Bilquis in Yemen in pre-Islamic times to dam wadi discharges from the mountains and to use them to feed an irrigation system around her desert capital at Marib. In Iraq on the Diyala Plains irrigation between the fifth and first millennium BC was achieved by breaching the level of the braided course of the Diyala river and by the construction of small distributive canals. Much later (around 600 AD) the Nahrawan canal was built in the same area to support a much more complex irrigation system. This canal was some 300 kilometres in length and fed a large integrated agricultural canal system. Developments of this kind have only been possible in the Arab world during phases of strong centralized government capable of organizing and maintaining extensive irrigation networks.

Two types of irrigation system remain common in the Arab world: flood irrigation, by which water is diverted regularly for a short period from a canal or river channel into a field surrounded by low earthern walls allowing water to sink into the soil, and furrow irrigation by which water is diverted to flow through furrows between the crops thus allowing soil water recharge in the root zone.

For rural dwellers in the Arab world inheritance of water rights has been as important as inheritance of land, since without access to water the land itself was useless. Water inheritance rules vary considerably between different hydrological regimes and between different cultural traditions. In the Middle Atlas of Morocco, for example, two types of inheritance system operate. One involves 'timed water'. Water is owned separately from plots of land and the owner is free to use the allotted time to irrigate land anywhere he or she wishes. Some water time is owned in blocks of as little as fifteen minutes. The second system involves sequential distribution of water with each piece of land being allocated a specific amount of water before the next plot is irrigated. Land and water in this system are therefore linked and cannot be separated for sale or other purposes (Geertz et al., 1979).

Over time, more and more ambitious schemes have been devised to use and re-use water in the Middle East. Syria's Lake Asad was built in the late 1970s to dam the Euphrates and was intended to provide water for 640,000 hectares of irrigated land. Foreign aid in Yemen has been used to build a new dam at Marib. The dam, completed in 1987, is designed to irrigate 20,000 hectares of former desert land when the distributive canals and associated irrigation systems have been completed (Figure 6.3).

All big dam projects require central organization and a level of agreement about how the water is to be distributed. Technological advances in the twentieth century have introduced the possibility of dams and irrigation systems on a larger scale than ever before, but it seems that advances in engineering have often outpaced organizational advances. Many dams have been built without adequate thought being given to how the dammed water is to be used, or to the effects on those living downstream of the dam who may lose out as a result of the extraction of the water from its former river course.

The 1970s in Morocco provide one example. This was a phase in

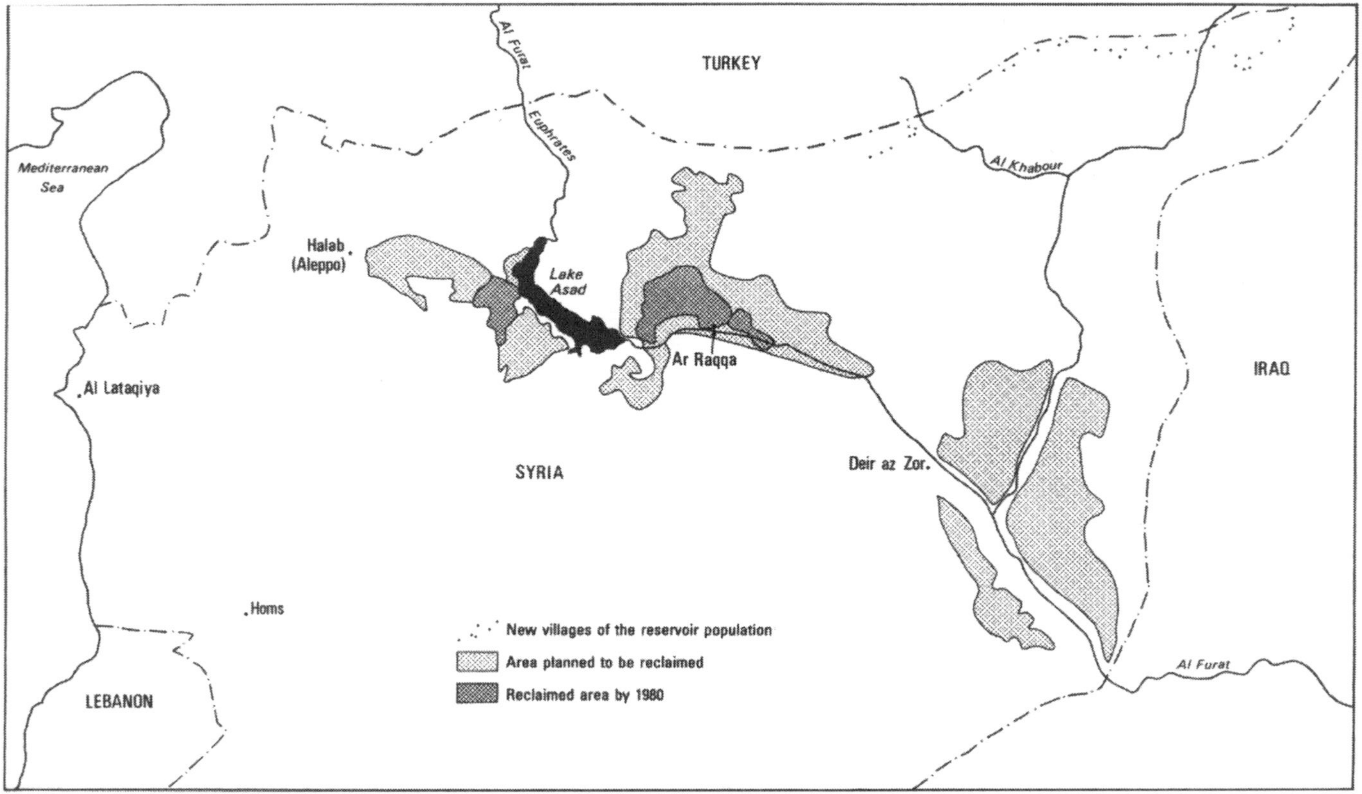
TURKEY
Al Furat
Euphrates
Mediterranean Sea
Halab (Aleppo)
Lake Asad
Ar Raqqa
Al Khabour
IRAQ
Al Lataqiya
SYRIA
Deir az Zor.
Homs
LEBANON
Al Furat
New villages of the reservoir population
Area planned to be reclaimed
Reclaimed area by 1980

Figure 6.3 Lake Asad and associated Syrian irrigation schemes

Moroccan development when the government decided to boost its earnings from agricultural exports and to reduce its dependence on certain imports such as sugar. As a result it chose in the 1973–7 National Plan to invest heavily in building large dams to extend the area of irrigated land in selected parts of the country. Table 6.1 shows for example that on the Gharb plain the intention was to more than double the irrigated area and in some places, such as the Souss plain, dams were intended to introduce substantial areas of irrigated agriculture for the first time. Table 6.1 shows that, despite building large dams, by the end of the plan period the area of land irrigated fell far short of the area planned. This was because of the improper phasing of the installation of water distribution networks from the large dams to the farms of the areas of privileged agricultural investment. This meant major delays in irrigating large parts of Morocco's best agricultural land, poor returns on government capital investment, lower agricultural productivity than could otherwise have been achieved and a loss of water from the dammed river systems to the disadvantage of other cultivators. For example the Idriss I dam completed in the Gharb basin in 1973 was designed to provide irrigation water to 125,000 hectares but by 1978 still only supplied water to 90,000 hectares. While the government of Morocco was apparently pre-occupied with financing large-scale irrigation projects to boost agriculture in those areas of the country oriented to the production of crops for export, it almost ignored areas of traditional rain-fed cultivation resulting in the area under wheat cultivation (a crop oriented to the domestic market) shrinking from 1,517,000 hectares in 1970 to 1,269,00 hectares by the end of the 1970s. Despite the problems and criticisms of the big dam schemes in Morocco, the government remained committed to this means of harnessing its water resources and in the 1980s launched the Mjara dam whose size is second only to the huge Aswan High dam in Egypt. Indices of agricultural production (discussed later in the chapter) quite correctly show considerable progress in terms of the increased value of Moroccan agriculture, but this has not necessarily meant an increased ability to meet the domestic food needs of the population since much of the increased agricultural value has come from exports of cash crops.

Large dams have been popular in many Arab countries because of their threefold promise: to increase the irrigated area; to provide potential for hydro-electric power and to control the potentially damaging effects of floods. They have been associated with a great many problems, however. In most cases the building of large dams

Table 6.1 The extension of irrigation in Morocco during the 1973–7 National Plan (hectares)

Region	Irrigated land 1972	Planned extension 1973–7	Actual extension 1973–7	Irrigated land 1978
Moulouya	38,330	13,590	10,005	48,335
Gharb	36,200	49,100	28,590	64,790
Doukkala	22,400	11,100	8,860	31,260
Haouz	8,500	18,500	19,500	28,000
Tadla	85,000	15,400	19,868	102,368
Souss	0	25,800	20,317	20,317
Loukkos	0	12,200	1,500	1,500

Source: Findlay (1984, 200)

necessitated the resettlement of rural populations. For example the flooding of the Euphrates valley to create the giant 625 kilometre square Lake Asad involved the resettlement of 60,000 people from their fields and villages. The Aswan High dam displaced 100,000 Nubians from inundated villages in the Nile valley. Dams also trap silt and reduce the potential of rivers to renew soil fertility downstream. Cultivators have therefore had to use more fertilizers, with all the consequences that this has for changing the aquatic life of the areas drained. In Egypt the use of nitrate and phosphate fertilizers quadrupled over the last twenty-five years and resulted in some severe aquatic weed problems, especially in the Nile delta.

Large dams also transfer the organization of irrigation to government institutions and away from the carefully evolved systems of rural society which over centuries dealt with the complicated issues surrounding water rights and the distribution of water. For example, all the farmers of the Nile delta and Nile valley are now ultimately dependent for their water supply on a small number of powerful centralized authorities. The Aswan High Dam Authority has complete power over water flow through the dam, while the Ministry for Irrigation determines how the water in the Nile is distributed through the many feeder irrigation channels in the valley and delta. The power of Middle Eastern governments to affect the course of rural development through the control of water resources has therefore never been greater. In the Middle Euphrates valley the rural population were led to believe that in the long run distribution of water from Lake Asad would improve their farms and their standard of living. In practice the farmers found that, as a pre-condition to benefiting from the new irrigation system, the government wanted them to consolidate their land holdings and to participate in a unified crop rotation. The government's motive was clearly to increase the efficiency of the agricultural system in which they were investing, but the changes were far from popular amongst the farming community.

The cost involved in building large dams has also necessitated that the water be redistributed largely to commercial farms and not to small peasant cultivators. The promise of receiving new irrigated land was one factor which initially helped to reduce out-migration from some parts of the Euphrates valley. For example at one point 7,500 farmers in the Balikh region were promised 5.3 hectares each, but revision of the plans, in the light of the costs of the project and the changing orientation of the Syrian government's agricultural programme, saw these criteria scrapped and the irrigated land given over instead to state farms (Meyer, 1990). The most serious losers, however, were the populations displaced by the creation of the dam. These people were forced to sacrifice their livelihood for what the government perceived to be in the national interest. The initial beneficiaries of the irrigation schemes on the Euphrates were the state co-operative farms of other parts of the Euphrates valley, while many of the displaced persons were resettled in villages with rain-fed agriculture along the Turkish border. As a result they faced a significant deterioration in their standard of living.

Other ecological effects noted in the case of the Nile valley, following the creation of the Aswan High dam, have included the decline of Mediterranean fisheries off the Nile delta because of the reduction in the nutrition levels of the water being discharged. There has also been some debate over whether the dam has resulted in an unproductive and unnecessary loss of water through evaporation – an estimated 10,000 million cubic metres of water are lost in this fashion each year (Gischler, 1979).

The damming of the Euphrates and Nile valleys have of course brought many benefits as well as problems. It is claimed that the Aswan High dam has made it possible to add over 20 per cent to Egypt's stock of irrigated land. Extension of irrigation has enabled the settlement of new areas, both in Upper Egypt and in other parts of the Nile valley. Groundwater levels around the Aswan High dam have also been raised, bringing potential for further agricultural development (see Case study I). Even in the previously well-established agricultural areas around the dam, the greater reliability of water supply has allowed up to three harvests per year to be achieved.

Case study I

Development in the desert: the Wadi Allaqi, Egypt
by J. Briggs

Faced with a population of over 50 million, growing at 2.5 per cent per annum, and a large balance of payments deficit, a good part of it related to the huge volume of food imports, Egypt has little choice but to explore all possibilities for increasing its agricultural output. As practically all of the Nile valley and delta already produces two crops per year, areas suitable for agriculture are being developed outside these traditional production zones. Currently, the government is undertaking a major land reclamation programme which has the goal of bringing into production over 3 million acres of land by AD 2000. The main targeted areas are around the delta and in selected oases in the Western Desert. Of lower priority, but with considerable potential, are the shorelands of Lake Nasser, created after the completion of the Aswan High dam in the mid-1960s.

Plate 6.4 Farmer with land enclosure in Wadi Allaqi, Egypt

Case study I (*continued*)

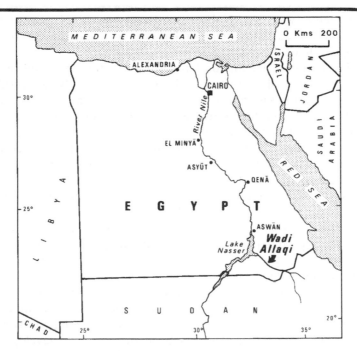

Figure 6.4 Location of Wadi Allaqi
Source: J. Briggs, University of Glasgow

There are several pilot projects in operation around the lake shore, the chief of which is located about 180 kilometres south of Aswan in Wadi Allaqi. The wadi is the main drainage channel from the Eastern Desert to the river Nile, even though it is completely dry at the surface for most of the time. The hydrogeology of the wadi and its region has been significantly modified since the formation of Lake Nasser. Indeed, some 100 kilometres of Wadi Allaqi were initially inundated (1978), followed by a retreat of some 20 kilometres along the wadi floor by 1991. Inundation resulted in lenses of shallow groundwater found at depths of only 15 metres, and at distances of 15 kilometres from the present lake shore. It is clear that inundation has had a significant impact on the much older, deeper-seated aquifers

Case study I (*continued*)

associated with the Nubian sandstone series of the region. Research is currently being undertaken to assess the nature and magnitude of these changes.

Furthermore, inundation has resulted in the deposition by the lake of high quality fertile silts. These are potentially very productive, at least in the short term. Research is currently being carried out to assess their long-term sustainability under different types of cropping and management regimes. The combination of these soils and the relatively shallow groundwater has led to the development of 20 kilometres of dense vegetation, predominantly tamarix, in the formerly inundated area of the wadi.

About 200 people have moved from the surrounding desert to the lower part of the wadi, close to the lake shore, over the last fifteen years. A four-element economy has developed. Charcoal production and livestock herding (sheep and camels) predominate, and in many ways are simply an extension or more secure adaptation of these people's former way of life. The collection of medicinal herbs for subsequent sale in Aswan forms an important source of income for some households. Completely new has been the adoption of cultivation in the wadi floor by most households. It is this that has been the focus of development interests in the wadi. People have been quick to try new crops and varieties. They have been prepared to experiment, and willing to try techniques new to them. A major problem is the shortage of labour, especially for irrigating the land. The problem is exacerbated by other demands on labour, such as for charcoal production and livestock herding. Cultivation is still only a minor activity.

Current development thinking on Allaqi embraces a number of ideas, both complementary and potentially conflicting. Resources could be placed in cultivation, especially in water pump technology, but this raises the question of the disposal of the extra produce. The nearest market at Aswan is 180 kilometres away. The returns may not be great enough to justify the levels of investment. Alternatively, large-scale farms have been suggested but the dislocation of the livelihoods of Allaqi residents, as well as the threat to this type of investment from future inundation, makes this a less than satisfactory proposal. Another option is to

Case study I (*continued*)

emphasize the management of existing vegetation resources in the wadi, and to complement the tamarix with the planting of acacia. This strategy would have two key implications. First the acacia would be highly suitable for charcoal production, and second it would produce grazing, thus improving the quality of local live-stock and that associated with camel trains in transit from Sudan to Daraw camel market (30 kilometres north of Aswan). There is a potential problem in that an environmental management struc-ture may be difficult to establish, especially among a settled Bedouin people who do not always perceive this kind of control as being in their interests. Overriding these issues are the political conflicts between, for instance, local and national interests, and between growth and conservation issues in development. Increas-ingly, private capital (provided by urban-based merchants) is also showing considerable interest in exploiting this new natural resource base. The Lake Nasser shorelands offer a great deal of potential, and Wadi Allaqi demonstrates in microcosm many of the debates surrounding agricultural development in the Arab world. In total, it is estimated that 300 square kilometres of land are available for development around the lake shore, but a clear understanding and considerable care are required to ensure that development is economically and environmentally sound.

There are many problems in terms of agricultural development which hinge on access to and use of water resources. The same is true for most other sectors of Arab economies. Increasingly governments must choose between water for agriculture and water for industrial and urban development. Exacerbating this problem is the fact that many countries in the region are competing for the same water resources and have based their development plans on assumptions which ignore the demands of other countries. The effect of Sudan's Gezira scheme on the discharge of the Nile has already been discussed. Ethiopia too has plans to divert the waters of the Blue Nile to improve its agricultural potential. Even more serious are the conflicting plans of Israel and Jordan over use of water in the Jordan river catchment area. Likewise Turkey, Syria and Iraq all have ambitious development plans for use of the waters of the Euphrates and Tigris. The problem may be illustrated by the following statistics. If the plans of all three countries for the Euphrates were fulfilled this would demand a flow of 36 million cubic metres per annum, while the recorded discharge at Hit in Iraq is only 31,000 million cubic metres. Consequently Iraqi farmers, who have drawn on the Euphrates for millennia to support their way of life, view with the utmost concern the steady lowering of the river's discharge following completion of the Keban dam in Turkey with its associated industrial/urban complex, as well as the irrigation schemes of the Middle Euphrates valley in Syria.

Capital, water and agriculture

For the oil-rich states the availability of large sums of capital from the 1970s onwards meant that many were tempted to seek new solutions to overcome the water constraint on agricultural development. Costly short-term solutions were found in mining deep aquifers and in installing desalination plants to convert sea water and brackish groundwater supplies for agricultural and industrial use. The sinking of deep wells in the Libyan and Saudi deserts has certainly produced startling results in the form of the patterns of crop production in the most arid of environments, as for example at Kufrah and Targa. Water mining however, has many of the same problems as large dams in terms of the levels of capital investment required and the need for expert knowledge to ensure appropriate irrigation techniques are used to prevent premature salination of soils. There is also the difficulty of deciding who should control and benefit from such schemes. In addition there is the very real problem that water extracted from aquifers takes centuries or millennia to recharge and, like all mining operations, resource exhaustion is ultimately inevitable. Water in deep aquifers should not be treated as a renewable resource.

The development issue which arises is not whether water should be mined, but whether its best use is for agricultural production and at what pace this precious resource should be depleted. (See Chapter 4 for further discussion of this issue.) It should be remembered that in Saudi Arabia groundwater supplies are being depleted far more rapidly than oil reserves. Many would argue that given the high cost of production and the impossibility of sustaining agriculture on the basis of tapping deep water aquifers, it would be better to use the water only for direct human consumption and other very high value purposes.

Table 6.2 Irrigation improvement targets, 1975–90 (areas in 1,000 hectares; costs in million US$)

	Cost	Area	Cost/area
Latin America	2,106	4,698	0.45
Asia	13,756	29,718	0.46
Africa	444	783	0.56
Middle East	6,263	9,789	0.64

Table 6.2 shows the costs of achieving the irrigation improvement targets of the Middle East and other developing areas up to 1990. It shows that the Middle East had by far the highest costs per unit area irrigated, and was also amongst those with the most ambitious targets for the extension of irrigation. While oil capital has made this more rapid pace of agricultural development possible, one must ask whether it is sustainable in the long run. For example in the United Arab Emirates surface water resources are almost non-existent. Oil capital was used in the 1970s and 1980s to sink many deep wells and to finance the development of new areas of irrigated agriculture. Impressive vegetable and crop production was achieved, but soon water table levels were found to be dropping and the incursion of sea water was recorded. It is only a matter of time before the costs of water extraction increase very substantially, the quality of water declines and the irrigated agricultural area begins to shrink once again.

Agricultural development schemes which rely on deep water mining depend effectively on 'fund' as opposed to 'flow' water resources. As a result they certainly do not offer the prospects for sustainable agricultural development, and must be questioned not so much on their high cost (although one might argue that the capital could be better deployed) but on the use of one of the region's scarcest resources for purposes which do not appear to yield enduring benefits to the Arab peoples.

Nomadism

'Pastoral nomadism is a livelihood form that is ecologically adjusted at a particular technological level to the utilization of marginal resources' (Johnson, 1969, 2). Given that 85 per cent of the surface area of the Arab world is too arid to support permanent rural settlement it is not surprising that nomadism has played a key role in the rural life of much of the region. Nomadism, however, is a human response to aridity which cannot exist in isolation from other forms of economic activity. Traditionally it has been viewed as interdependent with the two other main components of Middle Eastern society – the villager and the city dweller. The nomadic communities of the Arab region traded with the villagers and towns people to acquire certain food crops and artisanal goods. In turn they sold their camels as animals for use in settled agricultural work and for meat, and they raised revenues from transporting goods between urban centres across the most arid areas of the region. The terms of trade with settled communities varied over time, with the nomads often acting as war lords relative to the settled community, sometimes extracting rent in various forms from oasis peoples and from settled cultivators in the lands which fell within their domain. The twentieth century has totally ruptured these relationships.

The economic collapse of nomadism is captured succinctly in the following quote by the famous British traveller Wilfred Thesiger:

> Life in the desert ceased to be possible when the few, but entirely essential commodities that the Bedu had hitherto been able to buy in exchange for the products of the desert, became too expensive for them to afford, and when no one any longer required the things which they produced.
>
> (Thesiger, 1959, 95)

The extension of world trade patterns made it easier and cheaper to import meat from Australia and elsewhere. Tractors and pumps replaced camels in agricultural work, and in the age of the aircraft the deserts no longer presented the same great barriers to trade. The oases which the nomads had once controlled, and which had earned them an income through the sale of dates, now found their markets affected by date production in California and elsewhere. Settled cultivators found a rising demand for their products in the rapidly expanding cities, and with increased demand came small but significant increases in the prices of vegetables and cereals. Increased demand for food production also saw the extension of cultivation into areas where the nomads formerly grazed their animals, thus reducing their pastoral lands and producing conditions prone to cause over-grazing and soil erosion.

For nomads to maintain even a constant standard of living, it therefore became necessary either to increase their herd sizes (once again with the risk of over-grazing) or to pursue pastoral nomadism on a seasonal basis alongside other economic activities. The latter course has been by far the most common outcome. Table 6.3 shows how pastoral activities centred on the Saharan oasis of Kurudjel continues to form one part of the community's calendar. Nomad sedentarization in Kurudjel is usually dated from 1963 when the oasis became the site of a primary school and when some nomad families decided to settle for at least part of the year while pasturing their animals on surrounding grazings. In 1990 Kurudjel remained the pastoral base for flocks of camels and goats between July and November, but during the rest of the year the animals were taken away to find grazing elsewhere. As Table 6.3 shows, livestock movements are only one of several sources of income for the families of the oasis. Some family members work seasonally in the cultivation of grains and vegetables. Others supplement household income by artisanal activities, short-term labour migration or involvement in retailing.

Table 6.3 Calendar of activities in the oasis of Kurudjel, 1989–90

Activity	J	F	M	A	M	J	J	A	S	O	N	D
Agriculture												
Cereals						──	──	──	──	──	──	──
Vegetables	──	──	──									
Pastoralism												
Semi-nomadic		──	──	──	──							
Local pasture						──	──	──	──	──	──	──
Artisanal work	──	──	──	──								
Migration				──	──	──				──	──	
Commerce												
Vegetables					──	──				──	──	──
Dates							──	──	──			
Animals					──	──				──	──	
Crafts	──	──	──	──	──	──	──					

Source: Villasante-de Beauvais (1992)

Movement of pastoralists with their flocks continues to occur across large parts of the Arab world, but as a permanent livelihood form it has been greatly diminished. For example in Tunisia in 1860 there were estimated to be more nomads (600,000) than settled persons (500,000). By the 1960s the Tunisian censuses showed the survival of permanent nomadism as a livelihood form for only 4,000 persons, a number which had diminished yet further by the 1990s. In Saudi Arabia there were estimated to be still 3 million nomads on the eve of the Second World

War. Today they number only a few hundred thousand (perhaps 300,000) and the Saudi government has continued to encourage nomads to settle in the new areas of irrigated agriculture. This is a policy which seems to favour a switch from an age old livelihood form, well adapted to the ecology of the region, to a livelihood form which has been shown earlier in this chapter to rest upon the mining, over perhaps two decades, of a non-renewable resource. The Syrian constitution (Article 158) actually states that all nomads should sedentarize and by the time of the last census their numbers had dropped to a mere 20,000. In 1970 Mauritania and Sudan were the two countries of the region which had the highest proportion of nomads in their populations (45 per cent and 13 per cent respectively). In the last two decades Mauritania's nomads have shrunk to little more than 10 per cent of the total population.

Nomadic decline has resulted not only from changing economic circumstances but also from changing social and political environments. While in the past the nomads looked on themselves as a central element of Arab society, often leaving their families behind to collect dues from subservient peoples or to cultivate the oases, the twentieth century has marginalized their position. The authority of tribal leaders has been undermined in many key aspects. State authority, through the installation of national legal systems, has replaced judgements by tribal sheikhs. The drive to establish 'modern' education systems took young people away from herding activities and fixed them in one location to receive schooling. Above all, because of their inability to control wealth and opportunities for future generations, tribal leaders saw increasing numbers of young people leaving to seek urban employment or to work in other countries (rather than depending on patriarchal and tribal loyalty to achieve opportunities to earn a livelihood).

Of course in some areas contact with the world economy brought great power to a few nomadic families. For example, during the colonial era the sheikhs of the Gulf found that it was often advantageous to sign a peace treaty with a foreign power since this external recognition was often supported by the presence of foreign forces. This reduced the threat of challenges to power from other tribal leaders. The fossilization of tribal power structures which resulted inevitably favoured sedentarization by those in power, while the imposition and enforcement of state boundaries around the newly established sheikhdoms cut across old nomadic routes and discouraged the continuation of traditional patterns of pastoral nomadism. For example the Mutair used to move their herds from the head of the Gulf into the eastern Levant but found their grazing routes cut by the externally established boundaries between Kuwait, Saudi Arabia and Iraq.

Sedentarization

Active sedentarization policies have been pursued by several governments in the Middle East and North Africa. This policy has been favoured partly because nomads are perceived to be a difficult population group to administer and service. Many commentators suspect, however, that the more significant motivation underpinning the policy has been to establish a firmer control over nomadic groups, who are widely perceived as a political threat by those in government. This is not without some justification. The history of Morocco shows that many sultans were deposed and replaced by nomadic tribal leaders from the desert margins of the country, while relatively few successful uprisings against the sultan originated in the settled agricultural areas. Sedentarization has therefore been seen in some quarters as a means of reducing the independence of nomadic tribes and hence of neutralizing a politically volatile force.

The case of the Bedu of the Negev provides one example of a group which has faced on-going pressures from central government to sedentarize. These Bedu make up 13 per cent of the Arab population of the state of Israel. Throughout the earlier part of the twentieth century, and for the reasons listed above, many of the richer herd owners settled voluntarily. They were more able than others to buy land or buildings and to access the urban economy. The nomads who remained were therefore in general the poorer elements of the population, and less well able or less inclined towards sedentarization. The subsequent history of the Bedu peoples of the Negev has been researched in detail by the Palestinian geographer, Ghazi Falah (1989). Much of what follows is based on his many publications on the Bedu.

After creation of the state of Israel the Bedu were moved and restricted to a defined 'reservation' east of Beersheba. Eleven of the nineteen tribes were forced to leave their traditional lands. They were forbidden to leave the reservation except to attend the weekly market at Beersheba. New Jewish settlements were at this time being established in certain key locations in the Negev which had formerly been Bedu land, but the reserve also functioned to protect jobs in the Israeli labour market for new Israeli immigrants. Within the reservations some voluntary settlement took place, but most Bedu continued as best they could to follow their traditional lifestyles.

In 1966 the Israeli government decided to shift nomads to new areas and in particular to settle them in high density villages designated by the Israeli government. Settlement on their traditional lands was forbidden and rigorously controlled through a building permit system. This second phase sought to minimize further the amount of land occupied by the Bedu and was followed by attempts to persuade the Bedu to sell those lands which they could show were legally theirs. The Bedu were very resistant to concentration in the government settlements since this raised the prospect of them losing their lands through no longer living on them. However, the establishment by the Israelis of a so-called Green Patrol (from 1976), to locate and evacuate people considered to be illegally settled or using the lands of the Negev, ensured that the policy of shifting the Bedu continued unabated. In the 1980s further Bedu lands were expropriated to build three air bases. Some of the so-called illegal Bedu settlements were later recognized in an attempt to encourage voluntary sedentarization and population transfers to these locations. On other fronts the livelihood of the Bedu was further undermined – flocks had to be registered with the government or their grazing rights would be lost and the livestock seized by the Green Patrol. The grazing rights which were granted were often in areas with no adequate watering holes, making it impossible for Bedu to water their animals without leaving the areas where they were legally allowed to pasture them.

Given the continual pressure on the Bedu it is not surprising to find that they have become increasingly marginalized and impoverished. As a result of the events described above the Arab population of the Negev was still only at the same level in 1990 as in 1948. Over the same four decades the land under their control declined to a quarter of its former area.

In summary, the development or under-development of rural areas across the Arab world in the twentieth century seems to have been responsible not only for the emergence of some new types of agricultural activity, but also for the decline of other forms of rural livelihood, typified by the eclipse of nomadism. Some would suggest that the size of nomad populations has always risen and fallen in relation to the prospects and opportunities available to the settled populations of the region. Never before, however, have such strong forces combined to undermine nomadic life, and few would contest that once the traditions and skills of nomadism are totally lost from the younger generations they will be lost permanently, making any future return to pastoral nomadism in the region unlikely.

Land reform

It would be unfortunate if the impression was created that water resources were the only constraint on agricultural development in the Arab world. Truly remarkable feats in water engineering have been a characteristic of Arab lands for many millennia, and in recent decades the continuation of this tradition, in terms of new water development projects, has greatly increased the potential of agriculture. Other major constraints on agricultural development remain. Inequalities in land ownership and rigidities in land holding structures are one such obstacle to agricultural change.

In the early years after political independence some of the newly established regimes focused attention on the problems associated with the inequalities of land holding. A frequent contributory reason for the inequalities was the way in which rules about registration of land ownership had been introduced during the period of colonial rule. These rules were based on western concepts of private land ownership and ran contrary to the traditions of lands being held in common by tribal groups. The result was that in some places tribal land was registered nominally under the name of the tribal sheik. In other places tribal land was not registered at all, but instead was enclosed and claimed by urban merchants or by a small number of commercial farmers serving the urban market. In the Maghreb, French, Italian and Spanish colonists claimed large areas of land for their own agricultural settlement. After the end of the colonial era these states were faced with the problem of how to redistribute this land. The problem of land fragmentation was exacerbated in Islamic countries by the Koranic teaching that inherited private land should be divided equally between all male heirs. In some places the application of this ruling led to the branches of a single olive tree being owned by different persons.

Whatever the causes, by the middle of the twentieth century most Arab states had extreme inequality in land holding structures, with a few people owning vast areas and the majority of rural dwellers having only a few hectares or no land at all. For example in Egypt in 1952, 44 per cent of the rural population was landless, while 0.08 per cent of the population (2,000 land owners) held 20 per cent of the land. In Iraq

2 per cent of land owners had 68 per cent of the cultivable land, while in Lebanon even by the early 1960s the comparable statistics showed 5 per cent of land owners controlling 50 per cent of the cultivable land.

The result for agricultural development of the uneven pattern of land holding was that large land owners were more interested in producing cash crops (such as cotton in Egypt) to earn foreign capital than in establishing efficient forms of food production for the domestic market. For the small holder or landless labourer the problem was to produce enough food to meet the needs of his own family, once again resulting in minimal food production for the national market. Although it is dangerous to make generalizations, one might say that those with capital had no motivation to invest in food production while those with motivation had neither capital nor land.

The response from many of the new Arab governments in the 1950s and 1960s was to follow a socialist model of rural development, promising major land reforms. This was presented politically as a means of redistributing land to the poor and landless while at the same time serving to 'modernize' agriculture. In practice the two goals were seldom reconcilable. In Egypt and Iraq, land reform was one of the first measures introduced following military coups in the 1950s. In retrospect it would appear that the ultimate objective of these reforms was to destroy the power base and wealth of large land owners who had supported the previous monarchies. The temporary political union of Syria and Egypt saw Egyptian land reform exported to Syria. In Egypt the 1952 reforms set 84 hectares as the maximum land holding (later reduced to 21 hectares in 1969). During the Nasser period 12 per cent of the cultivable land was transferred to 10 per cent of the population. The redistribution of land and the consolidation of land holdings involved considerable work by the government land reform agency, while the operation of the new farms required the installation of a co-operative system of agriculture. Unfortunately the director of the government land reform agency often simply became the new absentee landlord. Those appointed as co-operative supervisors often had little agricultural knowledge and the control of credit and marketing of rural goods remained highly centralized and inefficient.

When the Ba'ath party came to power in Syria in 1963, it made new efforts to revitalize the land reform programme. The maximum size of rain-fed holdings was set at 300 hectares and a system of state co-operatives was established. This was a slow progress. Of the 1.2 million hectares confiscated, only around 700,000 hectares had been redistri-

buted by the end of the 1960s. Most land was placed in large co-
operatives and some was actually leased back to the original owners.
Only a comparatively small proportion was given over to landless
farmers in holdings of about 15 hectares on rain-fed land and 3 hectares
on irrigated land. But beneficiaries of this kind of scheme had to
participate in state supervised co-operatives. Many peasant cultivators
resisted this aspect of the reforms. Similarly in Tunisia, during its
socialist political phase in the 1960s, peasant farmers became disillusioned with the reform programme when they discovered that it
meant joining a co-operative system. Many threatened to slaughter their
livestock rather than give them over to state co-operatives.

In Algeria, land reform was desperately needed following the end of
the country's long and bitter struggle for independence. During the
colonial era the French had expropriated vast areas of the best land and
allocated it to French settlers, who turned the once fertile cereal-
growing areas into vineyards. The former Algerian cultivators of these
lands were forced to retreat south to less fertile and more arid regions,
triggering a process of environmental degradation which has continued
to the present day (Zaimeche and Sutton, 1990). Further population
upeavals and changes to the rural landscape resulted from the war of
independence (1954–61), during which the French 'regrouped' over 2
million rural Algerians in nucleated settlements supposedly to 'protect'
them from the FLN (Algerian liberation army). The war was also
responsible for the devastation of vast tracts of agriculture through the
use of napalm.

By the time of independence there was therefore a clear need for
major agricultural development, including land redistribution and other
reforms. In the first ten years, however, government attention centred
on the industrial sector since this was deemed the most appropriate way
to achieve national development. To quote Lawless (1984, 173) 'neither
the structure of ownership, nor the relations of production underwent
any substantial change'. In 1971 the government declared that an
agricultural revolution was to take place. Land from large and
absentee landlords was nationalized along with communal land. The
land was then redistributed and organized into 6,000 agricultural co-
operatives. One thousand 'socialist' villages were planned for the
'beneficiaries' of the reforms. But many who received land were
unenthusiastic about participating in a socialist co-operative production
system. Only a minority of the co-operatives proved to be commercially
viable and many initially lacked machinery, livestock and irrigation

equipment. Private land was not abolished, yet the government's socialist orientation meant that private cultivators received neither the means to achieve investment nor incentives to invest in more efficient systems. Private agricultural production therefore languished.

Land reform, while a necessary step to deal with the lands abandoned after independence, was not enough in itself to deal with the fundamental problems of Algerian agriculture – lack of enough cultivable land, shortage of capital for investment, and until the mid-1980s artificially low prices for agricultural produce (because of government policies to subsidize urban food). The result of these circumstances was that, twenty years after independence, Algeria had moved from being a largely self-supporting agricultural country to one requiring imports to meet more than half its food needs. Not surprisingly the 1980s brought a different approach to agricultural development. In the 1980s state farms continued to be heavily subsidized, but Algeria began to move away from the socialist ideals which had determined policy in the 1970s. Food costs were allowed to increase, allowing a 50 per cent rise in producers' prices in 1985 and 1986 alone. New land was sold off and loans were made available to private cultivators. Despite liberalization of agriculture and a movement away from the policies associated with land reform and Algeria's so-called agricultural revolution, the contribution of agriculture still only accounted for 9 per cent of GDP in 1988. Cereal imports to Algeria rose from 1.8 million tonnes in 1974 to over 6.1 million tonnes in 1988.

Land reform in the Arab world, and broader measures to intervene in agriculture for reasons of socialist ideology, have had a very chequered history. The uncertainty of rural restructuring made rural populations deeply suspicious of the motives of reformers. Programmes often proceeded too slowly, thus undermining the confidence of even those rural peoples who had something to gain. Co-operatives frequently proved inefficient because those involved either had inadequate knowledge of farming practices or saw no personal gain in working in state controlled agriculture. Land reform was also often attempted in isolation from wider agrarian reform, leaving rural communities without adequate marketing or credit facilities to reap the full benefits of new production systems.

While the land reform policies discussed above were far from successful, they showed at least some concern for rural development in the countries involved. In other countries such as Morocco no land reform was attempted, while in Oman the conservative political regime refrained more or less totally during the 1950s and 1960s from interven-ing in any way in rural development. In most Arab countries in the 1970s and 1980s agricultural development gradually shifted away from attempts at land reform and co-operative systems. Despite this change of emphasis there was little change in attitude and investment. Agri-culture continued to receive moderately little attention relative to pro-grammes for industrialization. In the Arab states only about 15 per cent of public investment in the 1970s was directed to agriculture. Instead national economic growth was assumed to flow from industrial develop-ment. Consequently 'industry' was seen to be 'modern', while agriculture was represented as 'backward' and undeserving of priority treatment.

Food security versus rural employment

Table 6.4 indicates the variable significance of the agricultural sector in the Arab world today. The contrasts between different states is stark.

Table 6.4 Indicators of the significance of agricultural systems

Country	Percentage of economically active labour force in agriculture			Percentage of arable land irrigated 1989[a]	Agriculture as percentage of GDP[b]		Per capita index of food production 1990[c]
	1975	1985	1990		1982	1988	
Algeria	39.2	27.6	24.4	4	6	9	94
Egypt	48.8	43.1	40.5	100	20	16	123
Iraq	37.1	24.1	20.5	47	7	15	92
Israel	7.9	5.1	4.3	49	5	n.a.	99
Jordan	19.0	7.7	5.8	15	7	7	108
Kuwait	1.8	1.8	n.a.	50	2	1	n.a.
Lebanon	17.0	11.3	8.8	29	n.a.	8	n.a.
Libya	23.5	15.6	13.7	11	2	4	102
Morocco	51.5	41.1	36.6	14	18	21	124
Oman	53.1	44.9	40.0	85	n.a.	3	n.a.
Saudi Arabia	56.3	43.7	39.0	37	1	3	n.a.
Syria	41.1	27.5	24.1	12	19	23	82
Tunisia	38.1	29.1	24.3	6	15	17	102
Yemen	65.1	58.6	55.6	21	26	28	79

Sources: FAO Yearbook (1991); World Bank (1990) *World Development Report 1990*
Notes: [a] Arable land is defined here as the area under arable and permanent crops.
[b] The data in this table are based on what are termed 'factor costs'. Recently the World Bank has also calculated the same index using purchaser prices. The result has been to boost the value of the agricultural sector relative to the figures quoted above.
[c] Based on the value of production per capita in 1990 relative to the average production value per caput for 1979–81.
n.a., not available.

The two neighbouring countries of Yemen and Saudi Arabia illustrate this well.

In North and South Yemen (united in May 1990) the majority of the economically active workforce are involved in agriculture (1.6 million of the 2.45 million labour force in 1990), yet as a sector agriculture contributes less than a third of the total value of the country's small GDP. A significant proportion of Yemeni agriculture continues to operate on traditional lines, with production being organized at a very local scale and with only a small fraction being commercial agriculture. Villagers produce for their own needs: government statistics suggest that the average farmer has only two cows, six sheep and four chickens! The 1980s brought vast changes to Yemeni agriculture on account of the capital influx from migrant remittances and the return in 1990 and 1991 of an estimated 1 million migrants in the wake of the Gulf crisis. Some remittances were directed into the purchase of tractors, which bestowed on the owners tremendous status but did not always benefit agricultural production. In some places tractors purchased by returnees were too powerful and, in the absence of training in ploughing practices, resulted in soil erosion. With no adequate employment opportunities outside the agricultural sector, and given the very limited skills of the majority of the population (see again Case study G), it is agriculture which must continue to provide a livelihood for the majority of Yemenis.

Ironically countries such as Saudi Arabia and Oman also show a high concentration of their labour forces in agriculture (about 40 per cent), but agriculture makes an insignificant contribution to their GDPs (about 3 per cent in both cases). The oil states inevitably find that all economic sectors are dwarfed by the scale of their oil industry and the marginal increase in the contribution of agriculture to GDP over the period mainly reflects the dropping value of oil products. Despite this, agriculture has been a recipient of capital investment in these economies, which employ a large proportion of their population in agriculture. Such investment has been one way in which national wealth has been distributed by those in control of the major revenues accruing from oil. In Saudi Arabia the dual goals of self-sufficiency in food and the fostering of agriculture as a source of jobs has been achieved partly by fixing generous farm subsidies. For example domestic wheat prices were at one point five times the world price and twice the cost of production. This use of capital to stimulate the commercialization of agriculture artificially and to raise productivity certainly gave the Kingdom one of the world's most dramatic rates of growth in agricultural production in the 1970s and 1980s, but it was an approach predicted on the availability of capital from oil revenues.

Most Arab states could not afford this. They had to choose between policies to increase the efficiency of agriculture in order to grow sufficient food to meet the demands of the rapidly growing urban populations or rural policies sympathetic to the needs and aspirations of their rural populations. Syria, Egypt, Tunisia and Morocco all had very significant rural populations in 1975 (between a third and a half of their active workforces were agriculturalists) and agriculture contributed between 15 and 20 per cent of national GDP. Algeria might also be included in this group of countries. Although agriculture contributes less than 10 per cent of Algeria's GDP, on account of Algeria's earnings from gas, the country had a very significant rural population and considerable agricultural potential. For all these countries mechanization, extended irrigation systems and improved food marketing structures offered ways of increasing food production in return for capital investment. Some or all of these measures were adopted in the poorer Arab states in the 1970s and 1980s, but most of these strategies involved sharp reductions in labour intensity, the re-organization of labour and land and the painful transformation of rural society. Table 6.4 shows that in all these countries agriculture shrank rapidly as an employer (down to under a quarter of the total workforce in Algeria, Tunisia and Syria). In Iraq, too, the decline in agriculture as a sector of employment was dramatic – from 37 per cent of all active persons in 1975 to only 20 per cent fifteen years later. These economies chose development paths which did not favour using the agricultural sector as a means of sustaining rural employment but opted for other goals. The resulting paradox emerges of an Arab agricultural system in which countries with the greatest long-term potential for agricultural development, in terms of renewable water resources and fertile soils, now have much lower percentages of their populations working in agriculture than do the more arid oil states such as Saudi Arabia and Oman.

In the absence of massive capital surpluses at the disposal of central government, economic growth in the agricultural sector in the poorer Arab countries in recent decades has taken precedence over other agricultural goals. Measures of economic growth in agriculture are difficult to achieve. One measure suggested by the Food and Agricultural Organization of the United Nations is the per capita index of food production value. Table 6.4 shows this index for certain Arab states. The index takes the average value of agricultural production per capita for the years 1979–81 as the base value of 100 for each country. Against this base the value of agricultural production has had to rise faster than population growth for the index to go above 100. By 1990, according to this measure, Morocco and Egypt with their ambitious irrigation expansion plans and using new high yielding grains were apparently making real progress towards raising the value of agricultural output. Having said this it is important to note that both countries started from a situation of massive food deficits and by the end of the 1980s were still importing very large amounts of food. In 1989 the food import bill for Egypt was still over US$5,000 million. Other countries such as Yemen faced major problems in restructuring their agricultural sector to feed their rapidly growing populations. In the case of Yemen an attempt has been made since 1990 to revitalize certain sectors of production such as fruit by banning imports and thus creating a protected sector for domestic producers. While this is possible for selected crops it is impossible to ban the import of staples on which the majority of the population depend. In addition the massive return of an estimated 1 million migrant workers to Yemen has greatly increased the potential rural workforce, undermining efforts to mechanize and modernize production.

The sustained growth of urban centres is possibly the one common force which historians will identify as critical in bringing about the commercialization of agriculture in the region in the second half of the twentieth century. There is evidence from many Arab countries of the role of urban capital in bringing about agricultural change as a result of the rapidly growing demands of the urban population for rural produce. In northern Syria, in the late 1940s and early 1950s, it was urban merchants from Aleppo who were responsible for the extension and development of cultivation (Meyer, 1990). They rented land in the Gezira from the nomadic and semi-nomadic tribal sheikhs who during the French mandate had been able to obtain the right of private property for their tribal lands. Using modern agricultural machinery cereal cultivation was established on large tracts of the best land. Later, irrigated cotton cultivation developed in line with the rise in world cotton prices. The urban merchants and later the tribal chiefs who also entered commercial farming benefited tremendously from this development of commercial cultivation while the sedenterizing nomads and villagers of the Euphrates valley gained very little or became landless labourers. It is not coincidental that the processes of agricultural change have often been associated with adverse changes in land holding patterns.

Syria is only one example from many across the Middle East of the role of urban merchants in the commercialization of agriculture. In Wadi Alliqi in Egypt (Case study I) it is urban-based merchants who have shown the greatest interest in buying up some of the improved lands on the wadi floor to develop them for crop cultivation. In Tunisia Signoles (1979) has shown how the diffusion of small vans amongst urban merchants was particularly important in accelerating the commercializing influence of urban merchants on agriculture. By extending their ability to reach more distant rural communities small pick-up trucks have facilitated the purchase of a wider range of rural produce for sale in the towns and made the production of marketable crops and vegetables more attractive to rural peoples wishing to sell produce and to buy urban commodities. The result has been a rapid change in the character of goods traded in rural markets and a shift away from cultivation for self-consumption towards the production of a higher proportion of rural produce for sale in the market.

A further factor affecting agriculture in the 1980s and 1990s is the changing character of urban food demand. As elements of the urban population have become wealthier, so demand for dairy produce has increased. Traditional agricultural systems were ill prepared to cope with these changes and extra food of this type had initially to be imported. The potential for domestic production often existed, but marketing mechanisms had to be developed to bring together domestic supply and demand. After a certain time lag this did happen for some commodities such as milk, through the creation with government assistance of marketing boards.

In large part, however, the agricultural sector has failed to attract public investment. Instead, many Arab states chose to depress urban food prices in order to keep down the cost of living for the growing populations of the cities. Subsidized food prices had a redistributive effect in favour of the urban poor, not at the expense of the urban rich but rather of the rural producer. In the 1980s the World Bank began to encourage countries to move away from this position despite widespread urban protests. Countries like Tunisia and Morocco began to privatize those elements of the marketing boards and organizations which they had set up. Early evidence of the results of this action suggests that so-called 'liberalization' has resulted in the transfer of control from a state monopoly to a small number of private owners.

One curious consequence of the neglect of agriculture in many of the poorer Arab countries has been that rural labour shortages emerged in the 1970s and 1980s. Considerable segments of the rural labour force responded to the very low level of agricultural wages by choosing to migrate (Lawless, 1987). For rural areas the consequences were severe. Agricultural production levels in many Arab countries continued to drop substantially behind population growth rates in the 1980s. With the exception of Saudi Arabia, the Arab states have become strongly dependent on imports to feed their rapidly growing populations. Between 1975 and 1985 the net deficit of Arab trade in agricultural produce grew from $400 million to $20,000 million. That the Arab region has become the world's largest food deficit area seems to be due as much to the under-development of its rural resources as to its annual 3 per cent growth in population.

Conclusion

The chapter opened by suggesting that agricultural development in particular countries needed to be judged relative to resource availability and development goals rather than against idealistic western models. In these terms it would be fair to say that some progress in food production was made during the 1980s through the use of higher yielding crops, the development of more sophisticated marketing systems (particularly in the dairy industry) and the extension of the irrigated area. The potential for further expansion of the irrigated area seems constrained by the finite water resources of the region while severe difficulties remain in increasing the efficiency of food production and at the same time seeking to sustain the traditional bases of rural society. The 1980s were a period when the agricultural sectors of most Arab countries failed to provide extra employment opportunities for the growing rural populations, consequently failing to halt the inexorable rural–urban migration of the population. This was as true in the heavily subsidized agricultural sectors of the oil economies as in the poorer Arab states (Table 6.4).

The oil-rich states may be able to sustain, for a little while longer, the exploitation of underground water reservoirs, in order to permit their capital intensive agricultural systems to prosper. Elsewhere liberal economic policies in most of the Arab states seem to place faith largely in market mechanisms, associated with the growing urban demand for a wide range of food products, to stimulate agricultural development. This process can only be expected to match the food demand resulting from the rapid rate of population growth if certain other steps are taken. First, Arab states will continue to find that their food needs can be met more 'cheaply' through imports from abroad, unless further investment in and re-organization of agriculture takes place. For urban capital to do this it may be necessary to privilege the domestic agricultural market over international market forces by protecting domestic production from less expensive foreign imports. Second, there is a need to ensure that rising urban food prices are passed on to the rural producer rather than allowing excessive profits to accrue to those in control of the food distribution system, including remaining state organizations. Only when rural wages rise significantly as a result of serving the urban market will rural producers become more motivated to increase the efficiency of commercial agriculture through better use of the human resources involved in agriculture.

These policies cannot be achieved easily either in political or economic terms. Without them agricultural change in the Arab world may continue to reflect forces favouring the 'under-development' of the rural economy, rather than the adoption of a sustained programme seeking to realize the considerable potential of wisely managed production systems based on the region's rural resources.

Key ideas

1 The history of agricultural development in the Middle East has been intimately linked with the evolution of increasingly sophisticated systems of water management. Progress associated with improved technologies related to storing and distributing water in arid environments has often been thwarted by inadequate attention to social and political matters, such as who is to benefit from new areas of irrigated land.
2 Extracting water from deep aquifers will permit only short-term agricultural gains, while water desalination schemes for agriculture are only viable in oil-rich states.
3 Powerful influences favouring voluntary sedentarization of nomads have been accompanied in many Arab countries by state policies to encourage or force permanent settlement of pastoralists.
4 Land reform was needed in many Arab countries to overcome the problems arising from inequalities in land holding patterns, but reform measures were largely unsuccessful because they did not encompass the wider conditions necessary to achieve rural development.
5 The search for food security is an important goal underpinning agricultural policies in many Arab states, but it is one which is often at odds with the need to sustain and enhance rural employment prospects.

7
Urban development

Introduction

The long history of urbanism in the Islamic world and the importance
of the inherited concept of the organic unity of urban Islamic society
have resulted in the emergence of geographically distinctive morphological, economic and social forms in the cities of the Middle East. This
distinctiveness has made them particularly interesting locations for the
study of trends and conflicts in economic development, since it has been
in the city that the most striking symbolic and material contests have
taken place between Islamic and non-Islamic forces in their separate
attempts to bring about change.

This chapter first considers the distinction between 'urbanism' and
'urbanization' as two different dimensions of Middle Eastern cities.
Second, it defines some of the morphological features of the Islamic
cities of the past which have survived and which infuse the contemporary urban environment with its culturally distinctive forms. This leads
to consideration of the forces which are responsible for bringing about
current change and development in the Arab city. Social scientists have
had to change their ideas on this topic rapidly over the last fifteen years,
switching from explanations of economic change induced dominantly by
external international forces to a recognition of the continued dynamic
of cultural forces within the Islamic world. The chapter concludes by
studying the opportunities offered by the economies of Arab cities to
their ever-growing populations as environments in which to live and
make a living.

Urbanism and urbanization in the Arab world

Significant population concentrations, dependent primarily on non-agricultural sources for their livelihood, have existed in the Jordan valley since at least 7000 BC. By the third millennium BC urban settlements are believed to have become widespread in Mesopotamia and the Nile valley. Each new civilization to sweep across the region produced either new urban forms or adaptations of pre-existing structures. It is not the purpose of this chapter to catalogue the fascinating findings of archaeologists with regard to the morphologies and functions of cities associated with the great civilizations of the past. These are mentioned here only to stress two points. First, it is clear that cities in the Middle East are not a new feature. Across many developing countries, the arrival of the colonial era marked the introduction of large urban settlements for virtually the first time, but in the Middle East 'urbanism' already had very deep roots. Indeed it is essential to recognize that the Middle East was the 'cradle of civilization', a term which reminds western societies that cities and many other symbols of western civilization owe much to their Middle Eastern origins. Second, it is evident from the succession of civilizations which infused Middle Eastern society with concepts of urban living (i.e. 'urbanism' as a way of life) that this cannot be studied meaningfully without also having an understanding of 'society' itself, and of the ways in which beliefs, culture, social institutions and political authority interact to make urban living possible.

If urbanism has been a long standing component of life in the Middle East, it was not until the twentieth century that urbanization transformed the region and boosted the populations of Arab cities. 'Urbanization' is often measured by the proportion of the population of a country living in large settlements. By this definition, urbanization advanced particularly rapidly from the 1960s onwards. In 1960 an estimated 32 per cent of the population lived in urban areas. By 1975 the proportion had risen to 44 per cent, while by 1990 the figure was 54 per cent for the Middle East and North Africa (58 per cent if one considers only the countries of the Middle East). For individual cities the rates of expansion have been dramatic. For example, Sanaa, the capital of Yemen, grew from 250,000 in 1980 to more than 660,000 in 1990, an increase of 164 per cent.

The continuing trend towards urbanization across the Arab world has implied a substantial shift in population from rural areas by the process of migration. Analysis of migration patterns across the Arab world indicates that some cities and some regions have gained much more population than others by the redistributive effect of the migration process. For example a comparison of the census data available for a wide range of Arab cities was undertaken by the author in the early 1980s. It showed that cities in Saudi Arabia and the oil states had grown much more rapidly because of migration than had the older urban centres of the Arab world. Typically, major cities in the oil states had over 70 per cent of their population consisting of in-migrants either from rural areas or small towns or international migrants. By contrast the equivalent statistics for North African cities ranged between 52 per cent for the population of the Moroccan city of Sale to a mere 11 per cent for Sfax, the second city of Tunisia.

Urbanization and urbanism are therefore different dimensions of Arab cities. Many urban dwellers in contemporary Arab cities are 'rural' in terms of their birth place, upbringing and lifestyles. Rural–urban migration has produced rapid urbanization without necessarily involving the adoption of urbanism. Inversely some of the oldest cities in the Arab world, despite being steeped in the traditions of urbanism, have been bypassed by the forces encouraging twentieth-century urbanization. This may well prove to have been a fortunate circumstance permitting the conservation of aspects of the urban morphology of certain ancient cities, but it has also meant that the meaning of Arab urban living has come to be contested. Inherited modes of urban life co-exist with rural values introduced through the migration process. These two traditions are also to be found contesting the meaning of urban space with the imported images of urban living which have reached the Arab city from so-called modernizing Western influences. All this raises the issue of what is meant by development in the Arab city. The question that must be asked of changes observed in the urban environment is 'Development: by whom? for whom?'

The Islamic city?

Western tourists often associate the cities of the Arab world with structures which long pre-date Islam, such as the pyramids and Sphinx at Giza, Cairo, or the amphitheatres of Carthage and El Djem in Tunisia. Traditions of urbanism in the Arab world for the inhabitants of its cities have little to do with these ancient relics and stem instead from a shared Islamic heritage. Islam has influenced virtually every aspect of Arab culture in terms of its influence on art, morals, law, customs, social behaviour and the structure of knowledge itself. The expression of cultural forces in the urban environment is evident through the ways in which *Sharia* law (the written rulings of Islam) has determined so many aspects of social behaviour. A few examples may help to illustrate the strength of this influence in the structuring of the urban environment.

Bonine, an American geographer, has shown how the Islamic ruling that Muslims pray five times a day facing in the direction of the holy city of Mecca has had some influence on the orientation of Moroccan cities (Bonine, 1990). When the armies of Islam set out to spread the faith in the seventh and eighth centuries it was this holy direction, or *qibla*, which defined the orientation of newly built mosques. In settlements where the mosque was built first, then the newly established city could be expected to evolve a street pattern whose orientation would be determined by the line of the *qibla* wall of the mosque. From the eighteen Moroccan cities studied by Bonine it seems that the *qibla* direction determined the street pattern and city axis in those cases where slope conditions allowed it. Of course the accuracy with which *qibla* directions were calculated varied over time and the relationship does not hold in the many cities across the Arab world whose origins pre-date Islam.

A second example relates to the organization of space within the Islamic city. The famous Arab explorer Ibn Battuta provides an account of the patterning of many cities in the fourteenth century. The striking feature of the Islamic cities, as opposed to the Chinese and Indian cities visited by him, was the influence of the mosque on surrounding land use. For many of the cities of the Middle East he describes the character of the *suqs* (bazaars) leading from the central mosque to the main gates of the city, thus establishing an image of the centrality of the mosque and its influence on land use patterns. Ibn Battuta describes the proximity of the central mosque to a range of related religious functions, such as the koranic schools and law courts, and also to related artisanal and commercial functions such as the distribution of book binders, candle makers and incense sellers. Further from the mosque were other *suqs* such as those involving textiles, carpentry and leather merchants. Further away still, either because of their 'unclean' nature or because of their inherent linkages with rural production systems, were the food suqs, the basket makers and the sellers of wool. These spatial relationships arose in part because of the ownership of significant portions of urban land by religious institutions (known as *waqf* land) and the involvement of the religious community in the organization of some craft and retail functions (Bonine, 1987).

Ibn Battuta also describes the separation of residential space in relation to religious observance for those towns on his travels in which the population was not entirely Muslim. In some cases religious quarters were separated by walls from the rest of the city and gates to these quarters closed at night and during Friday prayers. Segregation was evident in death as in life with separate cemeteries for Muslims, Jews and Christians around the outside of the walls of many Arab cities. Segregation of space in relation to the religious functions of the city is therefore one manifestation of the role of Islam in affecting the urban environment.

Orientalists have devoted much attention to describing the important institutions of the Islamic city. Undoubtedly Islam brought to cities in the Arab world many features which are not found outside the Islamic realm. Consider for example the *hammam* or public baths. Of course the Romans built many baths in their cities, but under Islam the *hammam* became an essential amenity to permit ritual ablutions and cleansing. Through time it became of much more than religious and hygienic significance. The *hammam* evolved as a social institution: a public space where certain agreed social behaviour could take place such as informal business conversations between men – a place for relaxation. Islamic society demanded either that separate *hammams* were built for men and women, or that separate times were defined for the use of *hammam*, in order to sustain the seclusion of women from men while in 'public'.

These three examples serve to indicate how the shared heritage of the Islamic faith strongly moulded the urban environment of cities in the Arab region. Some would go so far as to suggest that Islam is in essence an urban religion since only in a town or city can a person truly keep all the requirements of the *Sharia*. This suggestion has given rise to the view that it should be possible to define the spatial, social and economic characteristics of the Islamic city.

Schematic land use models have been devised by Ismail (1972) and others, incorporating the elements of the Islamic city which have been described above. These have great value in identifying the distinctive morphological characteristics of the medieval Arab and Persian city, but such models also involve many dangers. First, they have difficulty in capturing the nature of the built environment in those cities which existed prior to Islam and which inherited a structure formed as a result of non-Islamic forces. Second, there is the problem that urban land use models of any city type may tend to encourage a static image of the city. In practice, medieval Arab cities experienced phases of both expansion and contraction, requiring modification of the physical structures of the city. For example, Cairo under the Ottomans showed a tendency towards physical shrinkage. More important than physical land use models in analysing the development of the Arab city is the search for an understanding of the meaning of urban places within Islam. Third, there is the danger that generalizations about the Islamic city emerge which have been constructed in relation to only one or two Arab cities. Were Ibn Battuta's North African origins a source of bias in the urban features which he chose to emphasize on his travels? Does Ismail's model adequately consider the urban traditions of Yemen? Should academic effort be more concerned, as Abu Lughod (1987) has argued, with 'de-constructing' false images of the Arab city, rather than with reinforcing western stereotypes? Fourth, the project of defining the Islamic city may in itself blinker the student of Arab cities from recognizing that many non-religious forces were active in moulding urban forms even during the golden era of Islamic civilization. Military and commercial functions vied with religious ones for pre-eminence in the organization of urban space. Around the beginning of the sixteenth century it seems that proximity to military strong points within the city became more important. In Cairo, Damascus and Aleppo there is evidence that new *suqs* evolved around their respective citadels and some old *suqs* eventually also moved there.

All this leads to the question of whether it is really meaningful to talk of an 'Islamic' city any more than of a Christian or Bhuddist city. Islam as a fundamental cultural influence permeates the Arab city, but the city had developed in response to many other forces as well. What is important in the context of this book is that the city has a long tradition in the Arab world and has played a central role in the development of Arab society as shown in the writings of the famous Arab philosopher, Ibn Khaldoun. The interpretation of Islam, and of the meaning of the built environment within Arab Islamic society, has changed over time and has done so not only in relation to economic forces but also in relation to cultural, political and social relations.

Images of the city

In the contemporary Arab city there exist at least three competing images of the essence of urban living:

1 an Islamic image derived from the long traditions of orderly urban living, with the use of public and private spaces being dictated by interpretation of the *Sharia*;
2 a 'modern' view with the city as created in the Western image of urban space – an arena moulded by the needs of the capital;
3 a built environment occupied by poor rural in-migrants to the city, unable for economic reasons either to sustain the traditional urban environment or to benefit from access to the 'modern' city.

Can and should the Arab city be allowed to develop in relation to one of these images, or should it remain a contested space being moulded and remoulded by whatever happens to be the most powerful force at work at any one time? Discussion now turns to considering each image in turn.

Plate 7.1 The abandoned city of Marib, Yemen

Islam and the contemporary city

Of greater importance than the debate over whether or not it is helpful to produce generalizations of the historical Islamic city is the analysis of the role of Islam in the development of contemporary cities. In one sense the historical city lives on in many modern Arab conurbations in the form of the oldest areas of these cities with their mosaic of streets, *suqs*, citadels and mosques. The *medina* (Arabic for city) is a prominent and distinctive district of most cities in the Maghreb. In the Maghreb, particularly in Morocco and Tunisia, colonization led to the establishment of new cities built in parallel with the original settlements. The morphological contrast of the *medina* with other parts of the city centre of Maghreb cities is very stark. The colonists built districts in the style of their countries of origin. Colonial cities were often laid out on a grid iron pattern, as in Tunis and Sfax, featuring multi-storey apartment blocks constructed along broad avenues. Often a *Zone sanitaire* was left between the *medina* and the colonial city, but no significant space was left for the expansion of the old centres, creating problems for what would happen when their populations began to swell as a result of the joint effects of declining death rates and in-migration (Plate 7.2).

Plate 7.2 *Medina* and colonial city, Rabat

Plate 7.3 One of the many *suqs* of the *medina* of Tunis

In the Maghreb, the stark contrast of the *medina* and colonial town was no accident but arose from a colonial policy which encouraged separate development of the old Arab and the newer colonial cities. Fanon, the famous Algerian commentator and radical, describes the situation as follows:

> The European city is not a prolongation of the native city. The colonizers have not settled in the midst of the natives. They have surrounded the native City; they have laid siege to it. Every exit from the medina opens on enemy territory.
>
> (Fanon, 1967)

The modern cities built by the colonists became the seats of political and economic power, and within the city the role of the *medina* was challenged. While the religious functions remained, access was poor by contrast with that to the colonial town. As a result some commercial functions began to transfer out of the *medina*. Parts of the *medina* soon found their fabric being affected by attempts to instal new urban technologies such as electricity, gas and improved sewage systems. Other parts suffered from efforts to widen streets to allow access for the

Plate 7.4 An example of the street layout in the colonial-built city – Avenue Habib Bourguiba, Tunis

motor car. In many cities by the early twentieth century there were signs that the old urban elite and the richer merchant families were beginning to leave to live elsewhere in the city. They were replaced by large numbers of in-migrants unable to establish themselves elsewhere in the city. The situation in the *medina* of Tunis has been vividly described as follows:

> The refinements of a very calculated and controlled lifestyle have given way to all the components of slum living. . . . The in-migrant population is living catch as catch can, and has supplanted a

population which was steeped in urbanity; one that lived according to principles of order, privacy, and cleanliness of the various activities in space. . . . Now families live in each room of the old palaces and houses, while the former elite live in the European suburbs.

(Micaud, 1977, 145)

The picture presented above of the *medina* of Tunis might be extended to many of the other old urban cores of the Arab world. Decolonization in the second half of the twentieth century did little to halt the problems facing the *medina*. Departure of colonists from the 'modern' cities which they had built meant an accelerated transfer of Arab controlled administrative and commercial functions out of the *medina*, and the establishment of the 'modern' centre as a residential environment for Arabs working in the professions, administrative jobs and other parts of the 'modern' tertiary sector. Parts of the *medina* (particularly ethnic quarters left vacant by the departure of the Jewish community), along with large numbers of peripheral squatter settlements, absorbed poorer in-migrants to the city. With the rural economies of many of the Arab states in crisis, the flows of in-migration to the city tended to increase over time, creating severe problems in housing the growing population of migrants who arrived in the city without jobs or homes.

Detailed research using census material and house to house interviews has shown that the role of the *medina* in this process was a complex one. For example in the *medina* of Tunis by the late 1960s 76 per cent of residents were in-migrants and of these two-thirds had rural origins. There were parts of the *medina*, however, where the old large courtyard houses remained intact and continued to house the residue of the former *medina* population. Equally some sectors of the modern city centre, particularly those proximate to the *medina* formerly occupied by the Maltese and Italians, became distinct from the rest of the former colonial city centre in terms of the lower social status of the incoming poorer migrants. While decolonization did not reverse the fortunes of the *medina*, the sharp social divides between the *medina* and former colonial city centre have become more blurred.

Discussion in the previous few paragraphs has focused on the *medina*. Other areas of the Arab city might equally have been chosen. The purpose of investigating one part of the city in a little more depth has been to illustrate the way in which the image of one part of the urban environment has changed through time. The image of the *medina* as a distinctly Islamic part of the Arab city has been strongly contested.

The influence of Islam on the contemporary Arab city is not constrained, however, to the *medina* in North African settlements or to the
older cities of other parts of the Arab world. Evidence of the transfer
of the cultural symbols of Islam to the colonial built areas of the cities
of North Africa is evident even at the most obvious of levels in terms
of the many mosques which have been built in these districts in the latter
half of the twentieth century. Equally the transfer of some of the
commercial functions of the *medina* to the former colonial districts did
not necessarily mean an abandonment of the *suq* economy, but in some
cases meant the transfer of traditional forms of retailing and of retail
structure to a new location.

Even in those cities which have no *medina* (i.e. the majority of urban
settlements in the Levant and Saudi Arabia), the cultural influence of
Islam on the built environment is pervasive. The growth of Riyadh has
been strongly guided by Islamic principles. Organization of the central
areas of Riyadh reflects an attempt by city planners and politicians to stress
the city as a physical symbol representing Islamic values. In the oil states
the rapid growth of cities, in circumstances where finance has been
abundantly available, has meant that many new public buildings (banks,
airports, hospitals) have been constructed along the bold new designs
of so-called Islamic architecture. In all Arab cities the influence of Islam
on the physical environment is particularly evident in retailing practices
– for example the operation of *purdah* or female seclusion has affected
both the layout of shops and many aspects of retail behaviour. Men are
much more likely than in the West to undertake shopping for a wide range
of goods, and most sales staff are male rather than female. In Saudi Arabia
purdah has also meant the evolution of a dual public transport system,
with one set of buses operating for men and another for women.

Much more significant than the built environment is the way that
Islam has continued to influence social behaviour at a whole range of
levels. Table 7.1 for example shows the average number of hours

Table 7.1 Modal number of hours worked per day

	Monday	Tuesday	Wednesday	Thursday	Friday	Saturday	Sunday
Medina households	0	7	7	7	0	9	0
Households in former colonial town	8	8	8	8	8	4	0

Source: Findlay *et al.* (1984)

worked by people living in the *medina* of Rabat as well as in the former colonial central area. It shows that amongst those living in the *medina* work patterns continue to conform to the norms of the Islamic week with no work done on the holy day, Friday. By contrast occupants of the former colonial town seem to follow a European time schedule.

Since the Iranian revolution of 1979 the political influence of Islam on urban society has grown, making the city the locus for conflicts between those elements seeking to strengthen adherence to strict Islamic values and those parts of urban society more eager to 'modernize' and to align urban society with western values. The nature of these urban social conflicts is discussed later in this chapter.

The city as an arena moulded by the needs of capital

It would be possible to consider how capital, and in particular international capital, has affected the morphology of different areas of the Arab city by adopting an approach similar to that followed above with regard to the influence of Islam. Such an approach would no doubt show how the demands of capital through the processes of production, consumption and exchange required drastic changes in the urban morphology of Arab cities over time. Many old cities such as Tunis had their walls removed to ease traffic circulation. In Baghdad the Rusafah quarter was dissected by three 'modern' arteries cut through the old urban fabric. In Kuwait the occupants of the old centre were compensated for the loss of their land and rehoused in the suburbs. The old city was then totally reconstructed on Western rather than Islamic principles. But in creating new urban forms capital has followed different organizational paths.

What forms of production and consumption have emerged across the Arab world and how have these variations affected the development of different urban forms? Differences emerge not only at the level of urban morphology, but also at the scale of entire urban systems, comprising many cities linked to each other by a common political approach to production and consumption. Rather than being concerned with how capital has moulded particular parts of the city, this section of the chapter chooses instead to consider city systems as arenas of capital formation.

The sociologist Janet Abu Lughod (1984) has suggested that Arab cities can be classified in relation to their position in the organization of production and consumption. Four types of organization are identified below, following Abu Lughod's typology closely: neo-colonial systems, socialist states, charity states and oil and sand economies.

States such as Tunisia and Morocco have seen their cities evolve under conditions which can be described as neo-colonial. As the term suggests this implies a strong continuity with the colonial phase. While experiencing political independence, both Tunisia and Morocco have continued to be economic satellites of the West. Wealth continues to accrue in the 'modern' and largely foreign inspired sectors of the urban economy at the expense of the 'traditional' sectors. These countries continue to sell to the world market many of the same commodities as were important during the colonial era. They do so through similar marketing structures and often from the same colonial-built port cities. Dependence on foreign capital remains high given the policies favouring the establishment of 'offshore industrial production' in these economies. This has meant the creation of factories with foreign capital and under foreign control to produce commodities for re-export and not for the domestic market. These types of investment have been located in zones around the old colonial centres and have mainly 'benefited' the host countries through the absorption of some of their surplus female labour force. In terms of urban forms and processes, where international capital has come into conflict with other urban traditions it has nearly always been international capital which has triumphed. This resulted in the very rapid growth which took place in office construction in central Tunis in the 1970s and early 1980s, contrary to the stated objective of the city's master-plan to try to maintain the residential functions in the central area wherever possible.

State socialism has been the dominant political position of Algeria, Iraq and to a lesser extent Syria since independence. Clearly the realignment of states within the so-called New World Order which has emerged in the 1990s has affected the outlook of these states, but for most of the last thirty years they have followed economic courses based on state intervention, central planning and socialist ideals. State planning, while having many deficiencies, did have the avowed goal of seeking a more balanced regional pattern of development. Thus regional cities often grew in a more balanced fashion than might otherwise have been found, with secondary cities being sustained within the context of specific regional development projects. Unlike some parts of the Arab world, urbanization has been more strongly linked to economic development projects, rather than merely being a focus for new patterns of concentrated consumption. This is not to say that large squatter settlements do not exist around the cities of these countries, but for a time at least the attraction of primate cities was reduced by more forceful policies favouring a wider distribution of economic activities than would have occurred if socialist goals had not been sought.

A third set of cities identified by Abu Lughod were, perhaps provocatively, described as belonging to the 'charity' states. These include the cities of Jordan, Lebanon, Palestine/Israel and Egypt. The hallmark of recent urban development in these states has been their dependence on income from outside the state, both from foreign aid and from migrant remittances.

It has already been shown in Chapter 3 that Israel has received more in military aid from the United States than any other country in the post-war period with the exception of Vietnam. Between 1985 and 1990 the sum exceeded a billion dollars. Israel has also been the recipient of major donations from Israelis living in other parts of the world, in particular in the United States. In addition it has received massive loans and loan guarantees from the west to finance its 'development' strategies. Many Arab states have also been recipients of major financial aid. Table 7.2 shows the pattern of aid flows which occurred between Arab states up to 1989. For example, at the Baghdad summit of 1978 the Arab oil producers agreed to pay Jordan the equivalent of US$1,250 million per annum in recognition of its position as a frontline state *vis-à-vis* Israel and its costs as host to a very significant refugee population. Although this sum was in practice seldom paid in full, it represented a massive injection of capital since the GDP of Jordan in 1979 only amounted to US$2,269 million. In addition to Arab aid, Egypt and Jordan received significant levels of aid from the West. In the wake of the 1990–1 Gulf crisis patterns of Arab aid flows changed substantially.

How do aid and migrant remittances affect urban development? The most obvious result for cities of this form of dependent development is that it allows a physical separation of patterns of production from those of consumption. In most economic situations the city serves as a locus of both activities, since in most forms of economic activity employees and their families need to live near their source of employment. Consequently incomes are usually spent in the neighbourhood of points of production, but where income does not come from domestic production (i.e. in a so-called charity state) this relationship disappears. The city may become merely a location for consumption activities and those who administer the urban economy are freed in their decision-making from considering the usual balance of measurable costs and benefits. As a result cities such as Amman and Cairo have received extravagant investments in urban projects such as transport infrastructure. These cities also experienced considerable construction booms as migrants and their families invested remittances in housing and real estate (see again Case study H). Inversely the end of the affluent years in the oil economies saw remittances fall and many construction projects came to a halt half-completed. Remittances also served to fuel consumerism and boost imports of manufactured goods. This further fuelled expansion of the retail sector to a level which is out of balance with the productive urban base.

Table 7.2 Offical aid from Arab countries and agencies, 1973–89

Recipient	Amount (US$ million)	Percentage of Arab total
Syria	12,317	22
Egypt	9,364	17
Jordan	8,870	16
Morocco	4,345	8
Sudan	4,221	8
YAR + PDR Yemen	3,882	7
Other	–	22
Total aid to Arab countries	54,736	100

Source: adapted from International Monetary Fund (1991) *Financial Assistance from the Arab Countries and Arab Regional Institutions*

Abu Lughod did not include Yemen in this category, but given Yemen's immensely high dependence on remittances and aid one could claim that this country too has experienced a form of urbanization dominantly stimulated by the same forces as other charity states.

Oil and sand states are the fourth economic category defined by Abu Lughod. The nature of these economies has already been described in Chapters 3 and 4. Control of oil revenues by a very small number of Arab families created a curious situation where wealth was concentrated in the hands of a very small number of people. Some wealth was redistributed to citizens of the oil states through a welfare system which sought to provide health and education services, subsidized housing, pensions etc. The system also created a vast bureaucracy with well paid jobs for citizens of these states. Immigrant workers were largely excluded from these benefits.

The urban consequences of oil wealth was evident in the mushrooming of towns and cities in some of the most arid parts of the Arab world, since it was here that the families distributing the largesse had their palaces. Once again the cities which emerged were largely sites of consumption rather than production. These cities also became centres for massive immigration owing to dependence on foreign labour to construct and service the new urban infrastructure. Since migrants were not able to qualify for subsidized housing, and since most could not afford the expense of private housing (bought by increasing proportions of nationals), the result was the creation of highly segregated cities with high concentrations of different nationalities. This was particularly evident in Kuwait.

Abu Lughod's approach has several great strengths for someone wishing to study Arab cities in relation to issues of economic development. It helps to focus attention on economic processes rather than on urban physical form or geographical location. It shifts study away from consideration of individual cities to an informed view of cities within an interdependent international economic system. In particular it directs attention to how policy makers across the Arab world have evolved different policies affecting the mode of insertion of their economies into a wider global context, and the consequences of this for the cities of the states concerned.

While the approach is useful in pointing to the role of capital, as mediated through production and consumption processes, in explaining some of the variations in urban development, it is important to note the limitations of the approach. The earlier part of this chapter stressed the role of Islam in urban development. Clearly this and other aspects of culture are not directly addressed by the urban economy approach, because of the desire to relate urban processes to wider economic issues. While the analysis of capital relative to Arab urban development is a central theme it must not be allowed to eclipse the significance of culture, and few places in the world illustrate this as forcefully as the cities of the Arab world. Nor should economic processes be taken as the only international influence responsible for ordering Arab cities. International and national military pressures have also been critical and, amongst other forces, have been responsible for the large refugee populations of so many Arab cities with all the implications that this has had for their urban economies.

Home to the urban poor

The previous section has sought to examine aspects of the relation between capital and the city. Given the weak industrial base of most urban economies in the Arab states and the limited capacities of the formally organized aspects of the tertiary sector such as government bureaucracies to generate jobs, it is important to turn now to an examination of how those elements of the urban population who do not have access to secure employment manage to survive. Two interrelated aspects of survival for the urban poor are particularly important: employment and shelter. Poor in-migrants to the city arrive with little or no capital and with no rights to land to establish a dwelling. In order to survive they adopt what may be termed irregular or 'informal' means of finding work and shelter. These vary greatly from country to country, but some generalizations are possible. Attention turns first to how in-migrants search for work and seek to meet their needs through what has been termed the 'informal sector'.

The term 'informal' needs to be used with the greatest care, since it has been taken by some to suggest either that arrangements for work are disorganized or that there is complete freedom of entry to work in this sector of the labour market. Neither is the case. Informal activities such as hawking, street vending, brushing shoes, casual domestic work, construction labouring and so on share the common characteristic of requiring little capital or training on the part of the person seeking to earn a living in these ways. They may all be organized on a highly flexible basis in relation to the changing demands of the rest of the urban economy. So-called informal activities are often small-scale operations. Because of the abundance of labour they are usually labour intensive and often involve family labour. They provide a precarious existence with no guarantee of work or income from one day to the next. Given these characteristics it is not surprising that statistics on the importance of informal employment are hard to achieve and are at best somewhat unreliable.

One of the few attempts which have been made to carry out a comprehensive survey of informal sector employment across a whole country was undertaken in Tunisia in 1980 (Charmes, 1985). Despite the problems of producing statistics about this kind of activity, such as the difficulty of achieving workable definitions, of contacting and interviewing people employed on an irregular basis and of interpreting in a meaningful fashion the responses to such a survey, the results remain of considerable interest in helping to understand how the urban poor seek to make a living. The survey suggested that about 38 per cent of all non-agricultural employment in Tunisia was in small-scale informal activities (i.e. twice the number of persons employed directly by the Tunisian government). It emerged that the commonly held view of the informal sector as being dominantly concerned with commercial and service activities is false. Some 54 per cent of informal sector jobs were in small-scale manufacturing, particularly in textiles. Only a quarter were in commerce, the balance being in transport, hotels and catering and other services. Many people received their apprenticeship in the small-scale sector before going on to establish their own business on a more permanent and secure basis. Furthermore, small workshops of informal sector employees were often part of a larger network of workshops.

The survey concluded that the informal sector made a very substantial contribution to the urban economy of cities like Tunis, Sfax and Sousse, generating a substantial overall income (although offering relatively low pay to individual employees). The informal sector should not therefore be thought of as the end point of rural–urban migration in the sense of involving the transfer of rural under-employment to the cities, but rather as a productive and substantial contribution to the urban economy, albeit one that depends for its efficiency on the exploitation of apprentice and family labour.

In Cairo, as in many cities of the less developed world, domestic refuse from the wealthier parts of the city (and indeed waste from urban service locations and institutions such as hospitals and hotels) is collected by the workers in the informal sector rather than by the municipality (see Case study J). The German geographer Meyer has shown in his study of Cairo how 'formally' organized the 'informal' sector can actually be with the refuse collectors paying for the right to collect rubbish in a particular area (Meyer, 1987). This is feasible because the rubbish collectors make their living from recycling waste. Some of their income comes from feeding animals with the organic waste which they have collected, while other income comes from sorting and recycling waste paper, glass, plastics, fabrics, bones, footwear, tin cans and aluminium.

Case study J

A day in the life of a Cairo rubbish collector

Long before daybreak, the first of the approximately 2,000 donkey carts start their often one- to two-hour journey from the refuse settlements to the middle class and upper class residential districts of the Egyptian metropolis, in order to reach their working areas

Case study J (*continued*)

before the morning rush-hour traffic begins. When they arrive there, the small children look after the team of donkeys while their father or older brother starts collecting the refuse from the households. They have before them hours of going up and down stairs – often up to the sixth floor and higher – before they have filled their carts with the refuse from 100 to 250 dwellings.

Late in the morning, most of the teams are on their way home again. Behind, among the high pile of refuse, once can see not infrequently the dirty face of a sleeping child, with mouth and eyes black with flies. The driver sitting at the front nods off again and again, even in the dense road traffic. The donkeys find their own way; but then it becomes critical on steep stretches and unmade roads leading to the refuse settlement, and here the two or three donkeys often can only be persuaded with brutal blows with a stick to continue pulling the heavy cart.

When they finally reach the home farmstead, the single-axled cart is pushed backwards through the entrance gate and tilted back. The evil-smelling load pours through the open tail-gate of

Plate 7.5 Young refuse collectors, Cairo

Case study J (*continued*)

the cart and into the inner yard. While some of the rubbish collectors set out once more on a second refuse-collecting trip, after a short midday-break, the main work of the older girls and the women begins. They sort the refuse into re-usable scrap materials and organic waste, which is fed to the pigs. Ultimately, only the no longer usable refuse components such as filth, rusty tin cans and innumerable plastic bags remain. The last mentioned are burnt on nearby incinerating sites, but also next to the farmstead in some settlements.

When, in the late afternoon, the refuse fire flares up on the roads, which are far too narrow, and dense pungent smoke darkens the sun for hours and makes breathing an agony, being in the settlements becomes almost unbearable. Then if stronger winds fan the refuse fire for some days the flames encroach again and again upon the pitiful huts. The fire is usually put out quickly with the help of neighbours: but there have also been catastrophic fires – as in the year 1976, for example, when a large part of the refuse district at the foot of the Muqattam mountains fell victim to the flames.

Source: reproduced by kind permission of G. Meyer (1987) Waste recycling as a livelihood in the informal sector, *Applied Geography and Development* 30, 78–94

The complexity of the refuse recycling economy is therefore considerable. It depends characteristically, not only on a household division of labour with each member of the family having a distinctive and integral role in the system, but also on an interdependence of specialist functions held by neighbouring refuse collectors. Thus while rubbish collectors may operate in geographically dispersed areas of the city, sorting and recycling occurs in the residential quarters of the city occupied by the rubbish collectors. This illustrates how the informal labour market and housing market may often be linked. Attempts by the Cairo municipality to replace informal sector rubbish collectors in middle and upper class areas by a state organized system failed on several counts, not least of which was the fact that it raised the costs fourfold while simultaneously reducing efficiency.

The 'informal' housing sector, like the labour market, is highly organized and provides an essential but flexible means of establishing shelter for the urban poor in the Arab world. Because occupation of the land on which housing is established is often in some sense illegal, residents in informal housing often suffer from considerable insecurity of tenure and face problems in achieving access to urban infrastructure. Far from all informal housing is of a low quality however, and it may involve quite substantial investments by the residents. Nevertheless such residents remain illegal settlers and part of the so-called informal sector simply because construction takes place without planning permission, property transfers are not registered, and zoning and building regulations are not adhered to. The threat of demolition or redevelopment is ever present.

A typical sequence in the development of an informal settlement on the edge of the Tunis conurbation in the 1970s and 1980s has been described by Vigier (1987). A plot of land is sold by a land owner either to an individual family or to a small developer, who subsequently subdivides. Initially a family will build a single room and fence off their plot. As the family saves money the house is gradually extended, either by the family itself or using informal sector labour. As the family grows in size so extra rooms are added, often on a second storey. Later the house may be linked illegally to the water or electricity network. A building's state of repair and level of completion will depend on the family's income and how long it has been since the informal settlement was established. Figure 7.1 shows the extent of informal housing around Cairo. Although informal housing areas are less well serviced than their inhabitants would wish, and although housing collapses are not infrequent because of the lack of adherence to building and safety regulations, the system is efficient in producing shelter for the less wealthy members of urban society in Arab cities. The tragic consequences of failing to heed building safety regulations became evident in 1992 when a modest eathquake hit Cairo. The death toll of around 450 people was far higher than would have been expected in a Western city experiencing a quake of similar strength, and was blamed in large part on the poor house building standards of the city.

Estimates for Tunis suggest that informal sector building accounted for 40 per cent of housing construction in the late 1970s and early 1980s, and this was provided at 40 per cent of the cost of legally constructed residential areas. El Kadi's (1987) estimates for Cairo suggest that, for the same time period, informal housing construction acccounted for 82

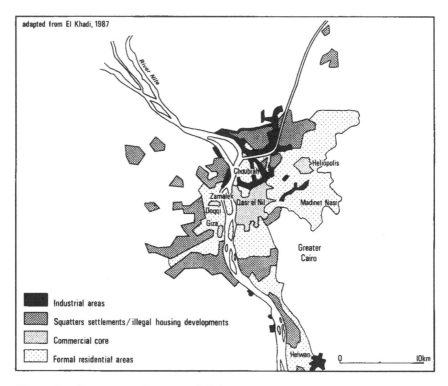

Figure 7.1 Squatter settlements of Cairo
Source: adapted from El Kadi (1987, 61)

per cent of new houses, and he predicted that this statistic would rise to 90 per cent by the year 2000.

While informal housing is a critical mechanism by which the poor gain access to shelter in Arab cities, it is also a means by which the rich get richer. This is so in several respects. For example, land is often sold in a highly organized fashion by land owners aware of the rising value of urban land. Alternatively land may be leased at high rents extorted in the knowledge of the insecurity of the illegal tenants. For urban planners informal housing provides a solution to the housing problems of the poor without the allocation of scarce funding to this important task. At the same time the illegal status of the settlements may provide an excuse for not providing urban services and infrastructures to these communities, while concentrating scarce resources elsewhere.

Arab banking as an indicator of urban economic development

One of the greatest problems in analysing Arab cities from whatever perspective is that the Islamic traditions of urbanism and the history of foreign economic forces have been so strong in moulding the urban environment that it is easy to ignore evidence of recent indigenous economic activity. This section of the chapter seeks to rectify this position. The evolution of Arab banking (as opposed to foreign banking in the Arab countries) will be used as a vehicle to illustrate the existence of a new type of economic hierarchy amongst Arab cities.

Within the Islamic world there could be said to be a cultural and religious ambivalence about the whole concept of any banking system based on the alien and anti-Islamic concept of interest payments (*riba*) on loans and deposits. The koran commands muslims not to take *riba*. Western influences within the Arab world have argued thst this term refers only to usurious interest for consumer credit. The stricter interpretation is that all interest, regardless of the rate and purpose, is contrary to the koran. Opposition thérefore to interest payments as a fundamental aspect of Western banking systems has been a severe dividing influence in the Islamic business community and ultimately led to the emergence of a complex and much debated set of financial procedures known as Islamic banking. Islamic banking is not discussed further here, but epitomizes one of the great problems for the region in becoming a centre of world banking. To achieve this status would involve bowing to culturally imported concepts which some would argue are irreconcilable with Islamic principles.

Figure 7.2 provides no more than a snapshot of the geography of Arab banking up to the early 1980s. It shows that Arab banking has mushroomed as a result of accruing oil revenues. There were as many new banks formed in the seven years 1974–80 as in the twenty-eight preceding years 1945–73. A second feature worth noting from the outset is the interesting hierarchical structure. Three cities – Kuwait, Beirut and Cairo – emerge as privileged locations, serving as headquarters for ten or more major Arab banks. They inevitably contest each other's claims to pre-eminence within the Arab urban system, but they also sit strategically within an Arab banking hierarchy which has extended its influence deeply into Europe as well as to other parts of the world economy.

Figure 7.2 suggests that there were at least three phases in the development of Arab banking prior to 1980.

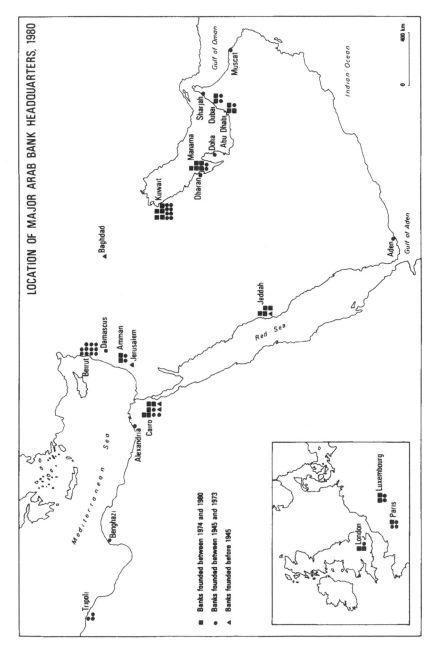

Figure 7.2 Location of major Arab bank headquarters, 1980

Arab banking before 1945

Foreign (i.e. non-Arab) banks financed most of the trade and development in the Arab world prior to 1945. There were some places, however, where colonial economic power was not absolute and where Arab capital began to organize itself independently. The financing of the Egyptian cotton industry was one notable example. The National Bank of Egypt, founded in 1898, lays claim to being the first 'modern' Arab bank. It was managed from Cairo, although involving substantial capital subscribed in London. In 1920 the National Bank of Egypt was joined by Banque Misr, establishing Cairo as the leading Arab centre of banking in the Middle East. Immigration from Europe stimulated Western style banking in Palestine, while the founding of the National Commercial Bank in Jeddah in 1938 coincided with the first receipt of oil revenues by the Kingdom of Saudi Arabia.

From the Second World War to the oil boom

In the early post Second World War years there was a rapid diffusion of Arab banking stimulated by several forces. The most significant of these was the granting of political independence. Amongst the first steps taken by many of the new Arab governments (Egypt, Syria, Iraq, Algeria, South Yemen and Libya) was the creation of Arab controlled financial institutions, the reduction or cancellation of the privileges of foreign banks, and the launching of national development plans which had to be underpinned by appropriate financial frameworks. For many states, political and economic independence was expressed by the adoption of socialist development policies. This meant a drive towards the nationalization of many different forms of private assets including banks.

If political independence was a major force in stimulating the spatial diffusion of Arab banking, then policies of nationalization were critical in producing a greater concentration of Arab banking in certain centres. Nationalization resulted in a flight of private capital from large parts of the Arab world. Beirut was the banking centre which attracted these funds more than any other location. It offered the potential of a location both for inter-Arab financial flows and for exchange between the Arab states and world economy.

The third factor which must be mentioned by way of explaining the expansion of Arab banking in the period 1943–73 is the growth of petro-capital. It is evident from Figure 7.2 that Arab banking in this period was not growing in proportion to the population, but rather was unevenly spread. There appears from Figure 7.2 to have been a particular growth of new headquarters in Kuwait, the cities of Saudi Arabia and the other oil-producing Gulf states. Oil revenues were therefore increasingly important as a catalyst to new banking activities.

The fourth formative influence in the period prior to 1974 was internationalization. Towards the end of the 1960s structural changes in the organization of Arab banking added to the search for investment opportunities for petro-capital and this encouraged the further internationalization of Arab banking. In particular a new generation of Arab–occidental banks began to emerge. These consortia (Arab capital and Western know-how) were advantageous to both partners in terms of the ability to tap into the growing commerce between Europe and the Arab countries. European financiers gained preferential access to petro-capital, and at the same time Arab banking systems were able to gain international experience and banking skills.

The years of affluence: 1974–80

When world oil prices soared, as a result of the deployment of the so-called 'oil weapon', this accelerated the process of capital accumulation in the oil-rich states. It also strengthened certain other financial flows which were to become significant for the Arab economies.

When oil revenues suddenly shot up after the 1973 oil crisis, Kuwait (unlike most other Arab countries) found itself with an already well developed international banking system. Increased budgets for national development and increased imports of consumer goods led naturally to an expansion of existing banking facilities and to the establishment of new banks. Jeddah and Riyadh also experienced a rapid growth in their banking systems. In addition the 1970s witnessed the substantial growth of banking in Manama (Bahrain). Manama emerged as a lower order centre for Arab banking for a number of reasons. Bahrain's oil reserves were small compared with those of other Gulf producers and were dwindling rapidly by the early 1970s. As a result the government sought economic diversification in a number of directions including the financial services sector. Unlike Kuwait foreign banks were allowed to operate within Bahrain. Manama became an even more attractive location for foreign banks after the introduction in 1975 of offshore banking legislation, making Bahrain like Singapore or the Cayman Islands. With the development of the international banking community it is not surprising that Manama also became attractive to Arab banks. It became the locus for some new inter-Arab consortia based in particular on co-operation in international banking between the oil states.

Outside the Arab oil states banking services grew throughout the 1970s. One of the driving forces behind the creation of new banks in these countries was the financial flows created by migrant workers sending home remittances to their families and relatives. New banks appeared in Egypt and Jordan in part related to this demand for banking services. In the meantime the continued growth of Arab banking around the world reflected the continued process of internationalization of Arab banking.

Figure 7.2 freezes the story of Arab banking in 1980. The position more than a decade later is very different and much more complex following the catastrophic crash of the Kuwait stock market, the severe shrinkage of oil revenues and petro-capital, the impacts on lenders and investors of the world debt crisis, the continued conflict in Lebanon, the evolution of the world banking system during extended years of world recession, the shakeout following the collapse of the Bank of Credit and Commerce International (BCCI), the increased strength and challenge of Islamic banking and last but not least the financial implications for the Arab banking community of the Gulf War.

The importance of Figure 7.2 is that it shows Arab capital as an active force in moulding its own urban hierarchies and economic systems. The positioning of Arab cities relative to each other and relative to the global economy may be interpreted in the light of the type of information presented above. What emerges from such an analysis is a picture of a financial hierarchy of cities which has evolved rapidly in relation to the changing imperatives of capital within the Arab banking system in order to find new and advantageous outlets for investment, controlled from a small number of Arab financial centres. Beirut, thwarted by the effects of the sustained conflict in Lebanon, has lost its position within the hierarchy to Kuwait. In the aftermath of the Iraqi invasion of Kuwait and the devastation of both the Iraqi and Kuwaiti states it will be interesting to observe how the Arab banking system re-establishes itself and in particular the effects on the Arab world's main banking city of Kuwait.

Discussion of the urban influence of the Arab banking system cannot close without stressing the obvious yet critical point that the urban hierarchy as measured by banking assets is grossly out of harmony with the demographic hierarchy of cities. Not only is the origin of much of the Arab world's wealth unevenly spread on account of the split between the oil-rich and the oil-poor states, but the operation of the Arab banking system has exacerbated this wealth divide. This has occurred through the concentration of private capital flows in the new banking cities of the region and the redistribution of petro-capital from the same banking cities to international investment opportunities often outside the Arab world and often serving the interests of Western economies. Arab banking cities have had a growing influence within the global economy, but this influence is not representative of the Arab world as a whole. The popularity of Islamic banking amongst some of those excluded from the 'modern' Arab banking system is not so surprising. In Arab cities, the mismatch between the needs of capital and the needs of the populations living in the urban environment has produced stark contrasts and inevitable tensions with respect to the ways in which the urban system is organized and has evolved.

As banking systems and financial flows become increasingly inter-nationalized, the potential for conflict within the urban arena and between cities with different ambitions within the financial system will grow. The ideological and military conflict between Baghdad and Kuwait (while having many other important dimensions) was but one extreme and tragic example of these tensions.

Conclusion: urban development and urban crises

This chapter has outlined some of the driving forces bringing change in the contemporary Arab city: (a) the ideological and cultural centrality of Islam. (b) the processes of capital in the production and reproduction of urban society and (c) the enormous explosion in the number of urban poor. These three forces are not independent of one another although each produces its own symbols in the built environment. These three, and other cultural and economic forces, contest the future of the city (Findlay and Paddison, 1984; Findlay, 1988). In some instances conflict-ing images of the urban future are resolved or accommodated, but in the 1980s and early 1990s the Arab city became increasingly a location of conflict. This reflected on the one hand a growing frustration in Arab society with the way in which state power was being manipulated and on the other hand an on-going crisis of identity. In the absence of alternative vehicles such as free and just democratic elections these crises erupted in a series of urban riots, strikes and demonstrations.

In the early 1980s there were food riots in Tunisia and Morocco involving the urban poor protesting against attempts to reduce or eliminate food subsidies. This economic policy had been introduced on the advice of the International Monetary Fund seeking to direct state development more rapidly along Western models. Cairo and Algiers also experienced urban riots associated with protests against falling living standards. The targets of the riots were urban symbols of the power structures upholding the state, such as local government agencies, and international hotels and restaurants. These riots did not seem to be orchestrated by Islamic militants or their supporters, but can most readily be interpreted as the urban poor protesting about the way that those in power have managed (mismanaged) the urban economy in the interests of international and domestic capital without adequate regard to the plight of the poor.

In the shadow of the Iranian revolution of 1978–9, Arab governments became particularly concerned lest urban riots signified religous unrest which might turn to revolution. Uprisings in Aleppo in 1980 and Hamah in 1981 did have Islamic political undertones and later problems in Algeria and Jordan were associated with support for Islamic move-ments, but overall there have been relatively few urban riots and protests linked with Islamic militants and some fourteen years after the Iranian revolution there has been no parallel event within the Arab world.

Since urban development and the growth of state power have taken place alongside one another, it is not so surprising that cities have become the focus of urban unrest and protest. It is in the cities of the Arab world that inequalities within society are most starkly displayed. It is in cities that the contested images of Arab society are most powerfully portrayed in terms of the symbols of the built environment. And it is in the urban riot or demonstration that those without power can most forcefully challenge political decision-makers about the future directions of state development policies. Development issues in the Arab world are therefore going to continue to be manifested and contested dominantly in the urban environment.

Key ideas

1 Islam has been a strong influence in structuring the physical, eco-nomic and social character of Arab cities. The concept of an 'Islamic city', however, is a construction of Western scholars and needs to be treated critically.
2 Several different paths to urban development can be identified in the Arab economies. Change in Arab cites may usefully be related to their evolving functions under different 'modes of production'.

3 The majority of city dwellers in the Arab world are poor, earning a living in insecure sections of the urban economy and living in low quality housing built illegally without planning permission.

4 Arab banks are one example of economic change in the urban arena resulting from Arab as opposed to foreign capital. They provide a basis for studying new international economic hierarchies emerging in Arab cities.

8
Arab identity and development

Introduction

What is Arab development? The chapters of this book have shown that the region has been an arena experiencing rapid and often painful economic, social and cultural transformations over recent decades. The changes which have been reported seem to fall into at least three categories, all of which could be described as 'Arab development' but which reflect different attitudes to and perceptions of the development process. These can be summarized as:

1 changes produced by *Western* intervention or *Western* interests in the region;
2 changes brought about as a reaction to and often rejection of Western values – this might be called *oppositionist* development since Arab values in themselves have not been the guiding force; instead development has been determined in a negative fashion in relation to external forces and has operated in an antipodean relationship to Western development;
3 changes arising from Arab interpretations of their own situation and development potential and which reflect progress and betterment of at least some Arab peoples in relation to their self-awareness of Arab identity.

Arab states have experienced changes stemming from all three types of influence. At any one time the forces may be operating in parallel within a country, producing contradictory if not conflicting effects.

THE TIMES WEDNESDAY SEPTEMBER 23 1992   27

On 23 September 1932, King Abdul Aziz Ibn Saud proclaimed a unified Kingdom in the Arabian peninsula. He called it, Kingdom of Saudi Arabia. This was the culmination of an endeavour that began in 1902, when he succeeded in re-establishing the rule of the House of Saud in the capital Riyadh.

The discovery of one of the world's biggest oil reserves in the Eastern Province of the Kingdom in 1938 helped to fuel the development of the country at an unprecedented rate. In particular during the last two decades Saudi Arabia has been transformed into one of the most prosperous and dynamic of world economies. Infrastructure and the welfare of the people have been developed to match the best anywhere. Yet, the Saudis have remained constant and true to their traditional values and Islamic beliefs. They have progressed beyond recognition, but have remained unchanged!

1932-1992
23rd SEPTEMBER

For further information please contact:
Foreign Information. Ministry of Information. Riyadh, Saudi Arabia, Telephone: +966 (0) 1 403 0701, Fax: +966 (0) 1 405 5218

Figure 8.1 Development without change?

Progress without change?

It has been argued in this book that historical circumstances inherited from the Western control of the region in the first half of the twentieth century greatly hindered the social and economic development of the Arab world (Chapter 2). The passing of the colonial era by no means brought Western influence to an end as is shown in other chapters of the book. Many of the changes subsequently brought about by contact with the West, however, were not desired by the majority of Arabs, nor could they be described as representing progress or development (Figure 8.1).

Changes which reflect a desire to move away from Western models of development in an 'oppositionist' stance are easy to find, since many of the Arab leaders of the immediate post-independence era adopted anti-Western public gestures in order to express their much acclaimed political independence. Classic examples discussed in this book are the policies of land reform introduced in Algeria, Syria and Egypt. Defiance of Western economic systems has also been shown in the history of the oil industry. Nationalization of Western oil company assets, for example, was a move which in itself proved of only limited benefit, since oil-producing states often remained dependent on Western technology and markets. Much more effective was communal action to develop, through OPEC, a mechanism of controlling oil production and supply to the world market. It is interesting that the most effective period of action by the OPEC states followed the identification by the Arab states of a clear and common objective for reducing oil supply (namely to bring leverage on the West to change its stance on Israel). One of the tragedies of Arab development is that there has been so little agreement on what constitutes worthy objectives for economic and social policy and action. Oppositionist stances to development, in the absence of a defined Arab identity, has tended to lead to abuse of the region's physical and human resources rather than to their effective deployment.

Examples of developments which are truly Arab, in the sense that they have arisen out of a firm Arab identity and reflect some conscious desire to move in an independent direction towards a better future, are hard to find. This is because there remains great difficulty in defining what is meant by 'Arab identity' in a way that is not merely oppositionist. So bitter has been the experience of contact with the West, particularly during the twentieth century, that most recent pan-Arab dialogue has been conducted in order to co-ordinate an Arab response to external threat or pressure. True attempts at independent co-operation between Arab states have a long history of failure. Consider for example the many vain efforts to forge political unions between Arab states and the relative failure of looser economic groupings between Arab states by comparison with the advances achieved over the last two decades within the EC. This brings discussion back almost to the point from which the book commenced. One of the reasons for failure in economic co-operation between Arab states is the very nature of the nationalism and the state. To quote from Mehmet:

> Nationalism . . . is a view of a world built on ethnicity and territoriality, ideas incompatible with Islamic universality. . . . The nation-state seeks to shift allegiance from God to the state. In return, it promises its citizens the benefits of socio-economic development in this life. Here too the nation-state conflicts with Islam. For it is precisely in the area of development that Islam has failed to evolve an ethos, to mobilize the masses for improving the quality of life. . . .
> (Mehmet, 1990, 1)

It would seem that political leaders in the Arab world either have had to focus the identity of their followers on the Western concept of the state or have had to pursue the universalist position of Islam. The latter position, as Mehmet notes, has a poor record in delivering what Western analysts would describe as socio-economic development. It has led instead to a rejection of the boundaries imposed by Western 'territoriality' and has contributed to the military conflicts which have so devastated the Arab region in the latter half of the twentieth century.

For those political leaders who have rooted their peoples' identity in national citizenship the record has been less bleak. This task was readily achieved in some places, such as Egypt, where physical and historical circumstances combined to heighten a local awareness. To encourage Arabs to view their primary identity as nationals of newly created states rather than as religious members of a greater community has not, however, been easy in any Arab country, and in many countries has met with a strong Islamic fundamentalist backlash. As the chapter on urban development showed, however, Islamic movements against national development have not thus far produced revolutionary conditions except in Iran.

The modest steps made by Egypt, Jordan and Tunisia towards introducing more accountable political leadership may serve in the long run to strengthen public identity with national development stategies and achievements. These countries cannot match the wealthier oil states, such as Saudi Arabia or the United Arab Emirates, on indicators of development such as GNP per capita, but in some senses they have moved much further than other Arab states by creating an environment in which national self-identity can emerge. This has partially been achieved through allowing more open discussion and debate of development objectives and paths. It is too soon to tell whether the promises of greater public participation and representation in government in other Arab states, in the wake of the Iraq–Kuwait conflict, will bring similar changes. Saudi Arabi in celebrating the sixtieth anniversary of becoming a unified kingdom insists that it has achieved '60 years of progress without change'. As tensions over its political and social structure mount, it remains to be seen whether or not economic development can really be achieved in the absence of substantial social reform. The evidence reported in this book suggests that the last two decades have brought dramatic economic changes to most Arab countries, but progress towards sustained economic and social development has been more limited.

Key idea

1 Conflicting interpretations of Arab identity exist. This remains a major constraint on coherent economic and social development in the Arab world.

References, further reading and review questions

Items for further introductory reading are indicated with an asterisk. Other items are either journal articles and books referred to in the text or material which the more advanced reader may wish to consult.

Preface

References and further reading

*Beaumont, P., Blake, G. and Wagstaff, J. M. (1988) *The Middle East*, London: Fulton.
*Chapman, G. and Baker, K. (eds) (1992) *The Changing Geography of Africa and the Middle East*, London: Routledge.

Chapter 1

Review questions

1 From Table 1.1 compare and contrast the key characteristics of Egypt and the United Arab Emirates, and attempt to describe the development issues which a Western and an Arab observer would identify in these two countries.
2 Try to explain why rising levels of gross national product per capita (Figure 1.1) are associated in a curvilinear fashion with rising life expectancies.
3 Why did Saddam Hussein describe the divisions of the Arab world as 'unnatural', and how does an understanding of his views on the Arab 'state' help towards a partial understanding of the Iraqi invasion of Kuwait in 1990?

4 Why has Islam been the focal point of those in opposition to state development policies in many Arab countries in the 1980s?

References and further reading

Ajami, F. (1982) *The Arab Predicament*, Cambridge: Cambridge University Press.

*Beaumont, P. (1989) *Environmental Management and Development in Drylands*, London: Routledge.

Drysdale, A. and Blake, G. (1985) *The Middle East and North Africa*, Oxford: Oxford University Press.

Gibb, A. (1940) *The Arabs*, Oxford: Clarendon.

*Hourani, A. (1990) *A History of the Arab Peoples*, London: Faber & Faber.

Hourani, A., Khoury, P. and Wilson, M. (eds) (1992) *The Modern Middle East*, London: Taurus.

McDowell, D. (1986) *Lebanon: a Conflict of Minorities*, London: Minority Rights Group (Report 61).

Peretz, D. (1978) *The Middle East Today*, New York: Holt.

Rowley, G. (1989) Lebanon: from change and turmoil to cantonization, *Focus* 39, 9–16.

*Salem-Murdock, M. and Horowitz, M. (eds) (1990) *Anthropology and Development in North Africa and the Middle East*, Boulder, CO: Westview Press.

Chapter 2

Review questions

1 What was the effect of the Ottoman empire on the subsequent political development of the Arab countries?
2 Explain the ways in which French colonization of Tunisia and Morocco changed the geographical structures of these countries.
3 Why is Saudi Arabia unusual in the Arab world in terms of its origins as a state, and what have been the consequences of these origins for the current character of the Kingdom?

References and further reading

Amin, S. (1970) *The Maghreb in the Modern World*, Harmondsworth: Penguin.

*Drysdale, A. and Blake, G. (1985) *The Middle East and North Africa*, Oxford: Oxford University Press.

Findlay, A. (1980) Patterns and processes of Tunisian migration, unpublished PhD thesis, University of Durham.

Hussain, A. (1990) *Western Conflict with Islam*, Leicester: Volcano.

Lawless, R. and Findlay, A. (eds) (1984) *North Africa*, London: Croom Helm.
Luciani, G. (ed.) (1990) *The Arab State*, London: Routledge.
*Robinson, F. (1982) *Atlas of the Islamic World since 1500*, London: Phaidon.
Rouissi, M. (1977) *Population et société au Maghreb*, Tunis: Ceres.
Weekes, R. (ed.) (1985) *Muslim Peoples*, London: Aldwych Press.

Chapter 3

Review questions

1 Comment on the development implications for Arab populations of the patterns of expenditure on military, education and health services shown in Table 3.1.
2 Describe briefly the history of Jewish rural settlement policy in the twentieth century, and explain why it has been a central issue in the Israeli–Palestinian dispute.
3 Why was the *intifadah* a necessary but insufficient condition for the creation of a Palestinian state?
4 Why has the absence of democratic elections been a hindrance to effective economic devlopment planning in many Arab countries?
5 What changes in industrial development have resulted from Egypt's *infitah* policy?

References and further reading

Boustani, R. and Fargues, P. (1990) *Atlas du Monde Arabe*, Paris: Bordas.
Coon, A. (1990) Development plans in the West Bank, *Geojournal* 21, 363–73.
Dimbelby, J. (1979) *The Palestinians*, London: Quartet.
Falah, G. (1989) Israelization of Palestine human geography, *Progress in Human Geography* 13, 535–50.
*Newman, D. (1991) *Population, Settlement and Conflict: Israel and the West Bank*, Cambridge: Cambridge University Press.
*Niblock, T. and Murphy, E. (eds) (1992) *Economic and Political Liberalisation in the Middle East*, London: Taurus.
Portugali, J. (1989) Nomad labour, *Transactions, Institute of British Geographers*, 14, 207–20.
Rowley, G. (1984) *Israel into Palestine*, London: Mansell.
Tripp, C. and Owen, R. (eds) (1989) *Egypt under Mubarak*, London: Routledge.
Usher, G. (1991) Children of Palestine, *Race and Class* 32, 1–18.

Chapter 4

Review questions

1 How has control of oil changed relative to the five main phases of oil production which the Arab world has experienced?

2 Why was Saudi Arabia described as a 'swing producer' within OPEC and why was this role one which Saudi Arabia was unwilling to sustain over the longer term?

3 In what ways could Libyan development planners be accused of confusing means and ends in the policies which they implemented in the 1970s and 1980s?

4 What kind of industrialization strategy has Saudi Arabia adopted and what are the underlying objectives of this strategy?

5 Critically evaluate the strategy of some Arab oil states which, in seeking to reduce their dependence on oil revenues, have become 'rentier states'.

References and further reading

Allan, J.A. (1985) Should Libyan agriculture absorb further development? In Buru, M., Ghanem, S. and McLachlan, K. (eds) *Planning and Development in Modern Libya*, 151–7, Wisbech: Menas Press.

Auty, R. (1988) The economic stimulus from resource-based industry in developing countries: Saudi Arabia and Bahrain, *Economic Geography* 64, 209–25.

*Bowen-Jones, H. (1984) The philosophy of infrastructural development. In El Azhari, M. (ed.) *The Impact of Oil Revenues on Arab Gulf Development*, 81–90, London: Croom Helm.

Chapman, G. and Baker, K. (eds) (1992) *The Changing Geography of Africa and the Middle East*, London: Routledge (read country-specific chapters on the Arab oil states).

*Drysdale, A. and Blake, G. (1985) *The Middle East and North Africa*, 313–45, Oxford: Oxford University Press.

Middle East and North Africa Yearbook, London: Europa (this yearbook always includes a chapter on oil and provides an accessible source on recent developments in the Arab oil industry).

O'Dell, P. (1981) *Oil and World Power* (6th edn), Harmondsworth: Penguin.

Unwin, T. (1988) Urban research agendas in the Gulf: the influence of the Saudi–Bahrain causeway, *URBAMA Fascicule de Recherches* 19, 37–54.

Walmsley, J. (1985) *Joint Ventures in the Kingdom of Saudi Arabia*, London: Graham & Trotman.

Chapter 5

Review questions

1 Why have the Arab labour-supplying states been unable to prevent a switch in demand in favour of immigrants from other sources?

2 Explain the housing and labour market implications, for the Arab labour-sending countries, of involvement with the international migration system.

3 Why is dependence on migrant remittances a highly vulnerable path to economic development?

References and further reading

Adams, R. (1991) The effects of international remittances on poverty, inequality, and development in rural Egypt, *International Food Policy Research Institute Research Report 86*, Washington, DC: IFPRI.

Al-Moosa, A. and McLachlan, K. (1985) *Immigrant Labour in Kuwait*, London: Croom Helm.

Birks, S. and Sinclair, C. (1989) Manpower and population evolution in the GCC and the Libyan Arab Jamahiriya, *World Employment Programme Working Paper 42*, Geneva: International Labour Office.

*Birks, S., Seccombe, I. and Sinclair, C. (1986) Migrant workers in the Arab Gulf, *International Migration Review* 20, 799–814.

Bourgey, A. (1984) Migrations et typologie des villes des Emirats Arabes du Golfe, *Etudes Méditerranéennes* 6, 225–44.

Findlay, A. (1987) The role of international migration in the transformation of an economy: the case of the Yemen Arab Republic, *World Employment Programme Working Paper 35*, Geneva: International Labour Office.

—— (1990) The changing role of women in the Islamic retail environment. In Findlay, A., Paddison, R. and Dawson, J. (eds) *Retailing Environments in Developing Countries*, 216–26, London: Routledge.

*Findlay, A. and Samha, M. (1986) Return migration and urban change. In King, R. (ed.) *Return Migration and Regional Economic Problems*, 171–84, London: Croom Helm.

Meyer, G. (1986) *Arbeitsemigration, binnenwanderungen und wirtschaftsentwicklung in der Arabischen Republik Jemen*, Wiesbaden: Reichert.

—— (1992) Labour migration into the Gulf region and the impact of the latest Gulf war, *Applied Geography and Development* 39, 106–25.

Seccombe, I. and Findlay, A. (1989) The consequences of temporary emigration and remittance expenditure from rural and urban settlements. In Appleyard, R. (ed.) *The Impact of International Migration on Developing Countries*, 109–28, Paris: OECD.

Chapter 6

Review questions

1 Why have large dam schemes brought as many problems as benefits to the peoples of the Arab world?
2 Explain the economic, social and political reasons for the decline of nomadism in the twentieth century.
3 Why have land reform measures met with so little success in the Arab world?
4 Describe and attempt to explain the agricultural trends evident in Table 6.4.

References and further reading

*Beaumont, P. and McLachlan, K. (eds) (1985) *Agricultural Development in the Middle East*, Chichester: Wiley.

Falah, G. (1989) Israeli state policy towards Beduin sedenterization in the Negev, *Journal of Palestinian Studies* 18 (2).

Findlay, A. (1984) The Moroccan economy in the 1970s. In Lawless, R. and Findlay, A. (eds) *North Africa*, London: Croom Helm.

Galaty, J. (1990) *The World of Pastoralism*, New York: Guildford Press.

Geertz, C., Geertz, H. and Rosen, L. (1979) *Meaning and Order in Moroccan Society*, Cambridge: Cambridge University Press.

*Gischler, G. (1979) *Water Resources in the Arab Middle East and North Africa*, Wisbech: Menas.

Ibn Battuta (1929) *Travels in Asia and Africa 1325–1354*, London: Routledge.

*Johnson, D. (1969) *The Nature of Nomadism*, Chicago, IL: Department of Geography, University of Chicago.

Lawless, R (1984) Algeria: the contradictions of rapid industrialisation. In Lawless, R. and Findlay, A. (eds) *North Africa*, 153–90, London: Croom Helm.

—— (ed.) (1987) *The Middle Eastern Village*, London: Croom Helm.

*Meyer, G. (1990) Rural development and migration in Northeast Syria. In Salem-Murdock, M. and Horowitz, M. (eds) *Anthropology and Development in North Africa and the Middle East*, 245–78, Boulder, CO: Westview Press.

Signoles, P. (1979) Mutations récentes des campagnes tunisiennes et intégration accrue à l'économie urbaine, *Urbama Fascicules de Recherche* 5, 143–86.

Thesiger, W. (1959) *Arabian Sands*, Harmondsworth: Penguin.

Villasante-de Beauvais, M. (1992) Quelques aspects sociaux et fonciers de l'oasis de Kurudjel, *Cahiers d'Urbama* 6, 67.

Zaimeche, S. and Sutton, K. (1990) The degradation of the environment through economic and social development in the 1980s, *Land Degradation and Rehabilitation in the 1980s* 2, 317–24.

Chapter 7

Review questions

1 Illustrate the distinction between the terms 'urbanism' and 'urbanization' with reference to the Arab world.
2 In what ways can the 'Islamic city' be considered to be a myth?
3 What does Abu Lughod mean by the term 'charity state' and how have economic circumstances moulded urbanization in the Arab 'charity states'?
4 Why have cities become the focus for unrest and protest in Arab countries in the 1980s and 1990s?

References and further reading

*Abu Lughod, J. (1984) Culture, modes of production and the changing nature of cities in the Arab world. In Agnew, J., Mercer, J. and Sopher, D. (eds) *The City in Cultural Context*, 94–119, Boston, MA: Allen & Unwin.

—— (1987) The Islamic city – historic myth, Islamic essence, and contemporary relevance, *International Journal of Middle Eastern Studies* 19, 177–204.

Bonine, M. (1987) Islam and commerce, *Erdkunde* 41, 182–96.

—— (1990) The sacred direction and city direction, *Muqarnas* 7, 50–72.

*Brown, K., Hourcade, B., Jole, M., Laiuzu, C., Sluglett, P. and Zubaida, S. (eds) (1989) *Urban Crises and Social Movements in the Middle East*, Paris: Harmattan.

Charmes, J. (1985) Development of the urban informal sector in Tunisia during the period of competitive growth. In Troin, J. F. (ed.) *Townsmen Cities, Urbanisation in the Arab World*, Tours: Urbama/Université de Tours.

El Kadi, G. (1987) L'urbanisation spontanée au Caire, *Urbama Fascicules de Recherche* 18.

Fanon, F. (1967) *A Dynamic Colonialism*, New York: Grove Press.

Findlay, A. (1988) International versus local economic forces, *Urbama Fascicules de Recherche* 19, 103–16.

*Findlay, A. and Paddison, R. (1984) *Planning the Arab City*, Oxford: Pergamon.

Findlay, A., Findlay, A. and Paddison, R. (1984) Maintaining the status quo, *Urban Studies* 21, 51.

Ismail, A. (1972) Origin, ideology and physical patterns of Arab urbanization, *Ekistics* 33, 113–23.

Meyer, G. (1987) Waste-recycling as a livelihood in the informal sector: the example of refuse collectors in Cairo, *Applied Geography and Development*, 30, 78–94.

Micaud, E. (1977) Urban planning in Tunis. In Stone, R. and Simmons, J. (eds) *Change in Tunisia*, 137–60, New York: State University of New York Press.

Vigier, F. (1987) *Housing in Tunis*, Cambridge, MA: Harvard University Press.

Chapter 8

Review question

1 In what ways have the ideals of state nationalism been incompatible with the concept of Islamic universality, and why has this been a problem for political leaders in the Arab world?

Reference

Mehmet, O. (1990) *Islamic Identity and Development*, London: Routledge.

Index

Numbers in *italic* refer to tables and figures

Aden 36
Agriculture 10, 27, 91–4, 113, 117, 119, 133–5, 142, *153*, 154, 158–9
Al Fateh 63
Al Hassan, Moulay 32
Ali, Mohammed 27
Algeria: agriculture *153*; employment 155; food production *153*; GNP 2; land reform 151–2; oil *2, 75*; population *2*
Algiers Summit 63
Allon Plan 55
Amman 59, 120–2, 174
Arab identity 3, 193; nationalism 11, 14, 19, 26, 42, 194; United Arab Republic 42
Arab League 20
Aramco 76, 84–6
Aswan High dam 135–7

Baghdad 10, 165, *184*, 188
Baghdad Summit 174
Bahrain 76; international migration *105*; Manama 186; ports 89; Saudi causeway 90, 100
Balfour Declaration 35, 38, 52–3
Banking 31, 183–8
Beirut 10, *184*, 188
Boundaries 2, 19, 35–43
British Petroleum 75, 78–9, 98

Cairo 10, 165, 174, 178–80, 181, *182*, 183–5

Camp David 68–9
Charity states 174
Cities *see* urbanization
Colonialism 26–34, *37*; administration 31; cities 167–8; British 26; French 27–31, 38; Italian 26; land ownership 32; Portuguese 32; settlement pattern 28, 32–3; Spanish 25–6
Cotton 156, 185

Dams 132, 133; Lake Asad 133, *134*; Aswan High dam 135–7; Marib 132–3; Nile barrage 27
Desalination 142
Development plans 68–9, 89–97, 109

Education 31, 34, 49, *50*, 90–1, 103, 105, 116, 119; literacy 111
Egypt: agriculture 27, *153*; Aswan dam 135–7; boundaries 41; Cairo 10, 165, 174, 178–80, 181, *182*, 183–5; colonization 26–7; education *50*; Fayoum depression 126, *127*; food production *153*, 156; foreign aid *175*; GNP 2; government 66, 68–9; health *50*; industrialization 27, 68–9; international migration 104, 116, 117, 123; land ownership 149; land reform 150; military expenditure 27, *50*; oil *2, 75*; population 2; tourism 68; Wadi Allaqi 138–41

Employment, agriculture 113, 117, *153*, 154, 158; informal sector 177–8; unemployment 106; urban 117; workforce 102, 104, 105, 108

Environment 7–11, *8*, 129–42; aridity 10, 129–44

Ethnic groups *12*, 14–15, 45; Armenians 14, 17, 36, 49, 52; Berbers 17, 49; Copts 13; Druze 13–15; Jews 35, 52–7; Kurds 17, 25, 36, 41, 49, 52, 65; Maronites 13–16, 35, 38; Nubians 49

Euphrates 7, 11, 136–7, 141–2

Food production 93–4, 99, 128, *153*, 155; food production value 155–6; imports 156; prices 157, 159, 188–9

Foreign aid 20, 52, 67, 174, *175*

Foreign debt 69–90

Foreign investment 69, 97–9, 185–9

Foreign relations 35, 52; military 51

Gaddafi 20, 42, 79, 92

Gaza 21, 55

GNP 1–6, 50, 195

Government 18–21, 46, 65–9; imams 111

Gulf states *see* individual countries

Gulf Cooperation Council 107

Gulf Crisis *see* Iraq–Kuwait war

Health facilities 5, 34, *50*, 90–1

Hormuz 9

House of Saud 39–41; King Abdul Aziz bin Saud 85; King Fahd 66

Housing 115–16, 118, 170, 176–8; squatter settlements 181–2

Hussein, Sadam 3, 54

Ibn Battuta 17, 130, 163, 165

Ibn Khaldoun 165

Industry, construction 68–9, 89, 91, 92, 95–7, 117

International Energy Authority 75, 88

Iran 13, 17, 18, 65, 78

Iraq 7, 10; agriculture *153*; Baghdad 10; boundaries 41; education *50*; employment 155; food production *153*; British mandate 38; health *50*; Iran–Iraq war 50, 80, 86, 97; international migration 123–4; land ownership 149; military expenditure *50*; nation state 3; oil *2*, 75, 76, 83; politics 48; population *2*; water resources 141–2

Iraq–Kuwait war 10, 21, 39, 50–1, 58–66, 82, 87, 100, 123

Irrigation 132–42, 143, 156, 158, flood irrigation 133; furrow irrigation 133

Islam *47*, 48, 172; Amal 18; banking 183; fundamentalism 13, 18, 19, 194; *Hizbollah* 18; Islamic Fundamentalist Group 19; land ownership 149; Metwali 13; Mohammed 47; purdah 171; Sharia law 19, 20, 25, 39, 40, 163–6; Sunni 13–15, 17, 38, 46; Shi'ite 13–15, 17, 38, 46, 65; Wahabbis 39–40

Islamic city 162–72; hammams 164; *medina* 167–72; mosques 163; *suqs* 163; *waqf* 163

Israel 52–9; agriculture *153*; Bedu 147–8; food production *153*; Gaza 21, 55; GNP *2*; Golan 55; Gush Emunim 56; Haifa 9; Jerusalem 56; Jewish National Fund 54; kibbutzim 54; Likud 56; moshavim 54; nation state 20–1, 53–4; Negev 147; population *2*, 57; settlement policy 54–7; Sinai 55; water resources 11, 58; West Bank 55–8, 62

Israeli–Palestinian conflict 20–1, 35, 52–64; Gaza 55; *intifadah* 63; Jerusalem 56; West Bank 55–8, 62

Jerusalem 56

Jordan: agriculture 117, *153*; Aqaba 9, 89; Amman 59, 120–2, 174; Baqaa Camp 59–62; education 50; employment 117; food production *153*; foreign aid 174, *175*; foreign relations 21; GNP *2*; government 66; health 50; international migration 104, 106, 107, 116, 117, *119*, 124; military expenditure *50*; Palestinians 58; population *2*, 161; remittances 102–3; urbanization 7, 174; water resources 11

Kurds 17, 25, 36, 41, 49, 52, 65

Kuwait 10, 100; agriculture *153*; banking 183–4, 186, 187; boundaries 41; education 50; food production *153*; foreign investment 97–9; GNP *2*; government 66; health 50; industrialization 97; international migration *105*, 106, 108, 116, 124; military expenditure *50*; oil *2*, 75, 76, 79, 83; population *2*

Kuwait Investment Company 97

Land ownership 32, 58, 149
Land reform 149–53, 193
Languages 16, 49; Berber 17; Kurdish 17
Lebanon 48, 187; agriculture *153*; Beirut
    16, 183–5; Beqaa Valley 13; civil war
    14–16; food production *153*; French
    control 38; foreign relations 35; Israeli
    invasion 52; land ownership 150;
    political conflict 13–16; population 2
Libya: agriculture 91–3, 142, *153*; Benghazi
    91, 92; boundaries 41; colonization 26;
    food production *153*; foreign relations
    10; GNP 2; industrialization 92;
    international migration 106–7; Jifara 91;
    Kufrah 142; oil 2, 75, 78, 79, 100;
    population 2, 91; rural–urban migration
    92; Targa 142; Tripoli 92; water
    resources 11, 91
Life expectancy 1, 2, 5–6
Livestock 140–1, 151

Mauritania: GNP 2; nomadism 145–6;
    population 2
McMahon 35, 36
Mecca 38, 40
Medina 38, 40; *see also* Islamic city
Migration, Asians 107, 115, 124;
    citizenship 115–16; international
    migration 51–2, 102–24, 176; policies
    109–10; remittances *105*, 118–19; rural–
    urban migration 31, 32, 158, 161–2,
    169–70
Military expenditure 45, 49, *50*
Mohammed, Prophet 11, 47
Morocco: agriculture 133–5, *153*; Atlas
    mountains 130; Casablanca 32, *33*; Ceuta
    32; colonialism 26, 31–3; education *50*;
    Fes 28, 31; food production *153*; foreign
    aid *175*; foreign relations 10; GNP 2;
    government 67; health *50*; international
    migration 119; Kenitra *33*; land
    ownership 32; irrigation 133–5, *136*;
    Marrakech 31, *33*; Meknes *33*; military
    expenditure *50*; population 2, *33*; Rabat
    32, *33*, 167; Safi 32; Tangiers 32, *33*;
    tourism 68; urbanization *33*; water
    resources 133, 135

Nasser, President 42, 150
Nationalism 194
Nation states 3–4, 19, 25–6, 36, 37, 38–41, 46
Nile 7, 161; barrage 27; Blue Nile *131*, 132,
    141; White Nile 130

Nomadism 39, 126, 128, 143–7; Bedu 144,
    147–8; Sanhaji 31; sedentarization 147–9

Oasis *145*, 146
Oil 9–10; boycott 78, 87; companies 75, 76,
    78, 81; exploration 76–8; prices 52,
    78–9, 81, 84, 86–8, 109; production 74,
    77–9, 82; refining 77–9; reserves 71–3;
    revenues 74, 83, 88–91, 175, 185, 187
Olives 126, 127
Oman: agriculture *153*; education *50*;
    employment 154, 155; food production
    153; GNP 2; health *50*; international
    migration *105*, 108, 109, 119; military
    expenditure *50*; oil 2, *75*; population 2
OPEC 75, *76*, 79–84, 86, 87, 193
Ottoman empire 23–6, 36, 165

Palestine 20–1, 52–9, 62–4
Palestinians 21, 38, 52–9; employment 58;
    intifadah 63; international migration 106;
    land ownership 58; population 57;
    refugees 16, 58–62, 64, 121
Phosphates 28, 32
PLO 16, 20, 63
Politics 18–21, 49
Population 2, 5–6, 33, 34, 57, 104; density
    129; rural 155; urban 2, 7, *33*
Ports 30–1, 89
Precipitation 7, 129, 130

Qat 111, 114
Qatar 75, 79; international migration *105*

Rabat 32, 33, *167*
Refugees 55, 58–62, *64*, 65, 121
Remittances 67, 102–3, *104*, 106, 117–18,
    119, 174
Rural society 31–2, 113, 140, 155; markets
    157

Sadat, Anwar 19, 68–9
Saudi Arabia 90; agriculture 93–4, *153*;
    Dammam 89; development policies 192;
    education *50*, 90; employment 96, 154;
    food production 93, *153*, 154; GNP 2;
    foreign relations 16, 20, 21; government
    66; health 50, 90; Hijaz 38, 40;
    industrialization 95–7; international
    migration *105*, 109, 124; Jeddah 184–5,
    186; Mecca 38, 40; Medina 38, 40;
    military expenditure *50*, 93; nation state
    38–41; national plans 109, 192;

nomadism 145; oil *2*, 73, *75*, 76, 82, 83, 84–6, 88; population *2*; Qassim 93; Riyadh 171, 186; Ruwala 126; women 19; Yanbu 95–6

Six Day War 54–5

Socialism 68, 151–2, 173

Standard of living 6, 98–9, 100

Sudan 49; boundaries 41; civil war 49; ethnic groups 49; foreign aid *175*; Gezira 141, 156; GNP *2*; nomadism 146; population *2*, 49

Suez 9, 27

Suqs 163, 165, *168*, 171

Sykes-Picot Agreement 35

Syria 7; agriculture *153*; Alawite 48; Aleppo 156, 165; boundaries 41; Damascus 10, 36, 165; education *50*; employment 155; food production *153*; foreign aid *175*; French control 38; GNP *2*; health *50*; Lake Assad *133*, 134, 136–7; land reform 150; military expenditure *50*; nation state 38; nomadism 146; population *2*

Tigris 7

Tourism 68, 162

Transport 9, 28–30, *29*, 77, 88–90

Treaty of Lausanne 36

Tunisia: agriculture *153*, 157; boundaries 41; colonialism 28–31; education 50; food production *153*; GNP *2*; government 66; health *50*; international migration 107; Kairouan 28; land reform 151; military expenditure *50*; nomadism 145; population *2*; Sahel 126; Sfax 30,

126; Sousse 30; tourism 68; Tunis 168–70, 173, 181

Turkey 11, 17, 23, 36, 65, 142

United Arab Emirates: Abu Dhabi 78, 108, 116; education *50*; foreign investment 98; Fujairah 89; GNP *2*; health *50*; international migration *105*, 123; military expenditure *50*; oil *2*, *75*, 78, 83; population *2*; ports 89; water resources 143

United Nations Works and Relief Agency 107

Urbanism 160–2

Urbanization 33, 116, 120–2, 156–7, 160–3, 172–82; population *2*, 7; riots 189; services 116, 172, 180, 182

Wadi 129, 138–41

Water resources 11, 129–42

West Bank 55–8, 62

Women 19, 97, 104, 123, 171

Yemen 42; agriculture 10, *153*, 154; education *50*; food production *153*; foreign aid *175*; GNP *2*; health *50*; international migration 104, 110–15, 117, 124; land use 129; Marib *166*; Marib dam 10, 133; military expenditure *50*; nation state 36, 38, 42; population *2*; remittances 102–3, *104*, 113–14, 154; Sanaa 161

Yom Kippur War 68, 79, 82

Zionism 20–1, 35